RELATIVITY AND COSMOLOGY

RELATIVITY AND COSMOLOGY

From First Principles to Interpretations

BALŠA TERZIĆ
Old Dominion University
Norfolk, VA, USA

Academic Press is an imprint of Elsevier
125 London Wall, London EC2Y 5AS, United Kingdom
525 B Street, Suite 1650, San Diego, CA 92101, United States
50 Hampshire Street, 5th Floor, Cambridge, MA 02139, United States

ISBN: 978-0-443-23542-9

For information on all Academic Press publications
visit our website at https://www.elsevier.com/books-and-journals

Publisher: Peter B. Linsley
Acquisitions Editor: Stephanie Cohen
Editorial Project Manager: Ellie Barnett
Publishing Services Manager: Shereen Jameel
Production Project Manager: Vishnu T. Jiji
Cover Designer: Miles Hitchen

Typeset by VTeX

Printed in USA

Last digit is the print number: 9 8 7 6 5 4 3 2 1

This book is dedicated to the memory of Henry Kandrup, my mentor and my inspiration. His brilliant thoughts are found throughout these lectures, keeping his memory alive.

Henry Kandrup (1955–2003)

Contents

Preface

This book originated from a set of lecture notes put together by the late great Henry Kandrup. As a first-year postdoc at the University of Florida, working with Henry on studying the role of chaos in galaxy evolution, I attended his graduate course in cosmology. Shy and seemingly timid by nature, Henry would come alive during his lectures – his animated gestures and his tongue-in-cheek humor, his charmingly rebellious disdain for authority and dogma, made every class a new masterful performance. Henry's profound analogies penetrated even the most impregnable topics; his uncanny ability to clearly, effortlessly communicate the mathematics of the Universe earned him the adulation of his students. We were all keenly aware that we were in the presence of greatness.

At the time of his sudden and premature death, Henry was in the process of turning his lecture notes into a textbook. Before passing, Henry shared with me an unfinished draft of his book, provisionally titled "Slightly Mathematical Cosmology," which is what this book is based on. I took it upon myself to bring his book to a level of completion where it can be shared with others. Books like this, on the subject as captivating and fundamental as the nature of the Universe, can never truly be finished; there will always be ways to explain a concept more clearly, or to cover an idea in a little more detail.

The aim of this book is to gently introduce a reader with no previous knowledge of relativity and cosmology to these exciting fields. But more importantly, this book is written in homage to Henry Kandrup – I wanted for his brilliant thoughts and his pedagogical insights to continue to teach curious minds, much as they did mine nearly 20 years ago.

This book is intended as an advanced undergraduate textbook. It spans almost exactly the amount of material that can be covered during a standard 15-week semester. After teaching an undergraduate course in relativity and cosmology from earlier versions of this textbook at Old Dominion University, I believe that it strikes a balance between the breadth and the depth of the material presented.

My contributions to this book roughly match those of Henry's. However, as you read this book, if something strikes you as particularly clever, the chances are those are Henry's words.

Belgrade, May 2022

Acknowledgments

I cannot help but marvel at the minor miracle that this book is being published! It has been my labor of love for over two decades, and would have remained an unpublished and perhaps forgotten manuscript had it not been for a number of people whom I acknowledge here.

This entire adventure begins with Henry Kandrup. He started writing this book, and I am honored to finish it. It is my hope that this book will keep his legacy and his memory alive.

I was about 8 years old when my father, Petar Terzić, took me to a public lecture about black holes. I was instantly hooked by their mystery, intrigued by their darkness. Growing up, my father was my most dominant role model, an unrelenting champion for science! He showed me by example what devotion to science looks like: staying up late, reading, studying, discussing physics.... Watching Carl Sagan's "Cosmos" further inspired me to look to the stars. I was fortunate that in Serbia, where I grew up, physics is introduced to the curriculum in middle school. There, I was influenced by another of my childhood heros, Nenad Golović. Nenad encouraged my interest in solving physics problems and introduced me to physics competitions. After that, my calling was cemented!

My doctoral advisor at Florida State University, Christopher Hunter, had a formative role in my development not only as a scientist but also as a person. From the beginning, he treated me as his equal, even though I never was and probably never will be his equal. This approach had a profound impact on my life in academia and beyond – I thought, "If this intellectual giant, this exceptional human being does not look down at me, no one should!" What a powerful boost that was! Chris's rigorous-yet-pragmatic approach to mathematics is something I have been trying to emulate. He formally introduced me to astrophysics, which became my first academic love.... And you always remember your first love!

Court Bohn, an astrophysicist-turned–accelerator physicist, convinced me after Henry's passing to join his Beam Physics and Astrophysics research group at Northern Illinois University. Court ushered me into the exciting field of accelerator physics, which has remained my home ever since. I learned a great deal attending his astrophysics class. Some of his thoughts and fastidious style of presentation made its way into this book.

My career in the sciences was salvaged in 2009 when Geoff Krafft offered me a staff research position at Jefferson Lab in Newport News, Virginia. By then, long years of postdoc "purgatory" have exhausted me to a point of interviewing for quant positions. Geoff stepped in, offered me a permanent position, and saved me from becoming filthy rich yet miserable. I have known since my sandbox days that science is my calling; anything else would be a miserable compromise for me. What ensued between Geoff and me was a profound scientific connection and the most prolific collaboration, which continues to this day. In his inimitable "teacher's teacher," ruminating style, his guest lectures in my relativity and cosmology class led me to new insights which found their way into this book.

Studying nonlinear effects of general relativity with Alex Deur allowed me to make an academic return to astrophysics and cosmology. He generously devoted many hours to meticulously reading through this manuscript. His many perceptive suggestions and comments significantly strengthened the final product.

My dear friend Anton Arkhipov provided the art on the cover of the book. The image originated from a photograph Anton took during our whirlwind trip to Tennessee to observe the 2017 solar eclipse – one of the most exhilarating experiences of my life. In keeping up with the deeply personal nature of this book, I am delighted that this unforgettable memory graces its cover.

The editorial team at Elsevier, Stephanie Cohen, Ellie Barnett and Vishnu Jiji, patiently shepherded this project to its publication. I am indebted to them and to the anonymous reviewers who provided invaluable feedback that improved clarity and completeness.

My home institution, Old Dominion University, generously supported my sabbatical in Belgrade in the spring and summer of 2022. Without that respite from faculty obligations, it is unlikely that this book would have come to publication.

I thank my students who read through the book and made valuable comments: Corey Sargent, Ali Mand, Jackson Clark, Penn Rogers, and Antonia Seifert.

I express gratitude for many years of insightful scientific conversations with Alister Graham, Ilya Pogorelov, Ily Vass, Ioannis Sideris, Dan Grubb, Slava Derbenev, FX Girod, Vasiliy Morozov, Tommy Michealides, Emery Conrad, and Alexander Conrad.

I will always be immeasurably indebted to Garrett Walsh and Carl Gay and their world-class team at MD Anderson in Houston, Texas. Without them, none of this happens.

My dear family, mother Slavka Terzić, father Petar Terzić, and sister Katarina Conrad, always unreservedly believed in me, often more than I believed in myself. When moments of doubt would cloud my judgment, my eternal, unassailable cheering section would provide me with an instant boost of confidence and ambition.

I am forever grateful to my wife, Denise Tombolato Terzić, for her steadfast support and generous sacrifice. Her timely encouragements and "gentle" nudges were instrumental in bringing this work to fruition. She made me realize that my notes were not mine to keep, and that I should share them with the world.

Finally, my beloved daughter Nina Terzić inspires me to be the best version of myself, every day. Making her proud of her daddy's accomplishment is the most important validation of my work.

Organization of the book

Big Picture: Before we get into the crux of the discussion, it is useful to preamble briefly the new material with the "big picture" view of its purpose and importance.

Notations and Notes: A separate box is designated for the notation adopted in the book, as well as various notes that are of importance for following the discussion in the text.

Examples: Worked-out examples are boxed in order to reinforce concepts presented in the text.

Most important *definitions* and concepts are enclosed in a separate box with the darkest background.

CHAPTER 1

Introduction

The most incomprehensible thing about the world is that it is comprehensible.

Albert Einstein

Astrophysics deals with the physics of the Universe, including the physical properties (luminosity, density, temperature, chemical structure) of celestial objects such as stars, galaxies, and the interstellar medium, as well as their interactions. Astrophysics is a very broad field: it includes mechanics, statistical mechanics, thermodynamics, electromagnetism, relativity, particle physics, high-energy physics, nuclear physics, engineering, and others.

Cosmology is astrophysics on the largest scale, where general relativity plays a major role. It deals with the Universe as a whole – its origin, distant past, evolution, structure. When looking at the Universe at such grand scales, the locally "flat" and "slow" approximation – the realm of the Newtonian mechanics – is no longer justified.

As its subject matter involves important and overarching questions, such as: 'How did we get here?', 'Was there a beginning?', 'Are we special?', thus heavily flirting with philosophy and theology, modern cosmology has proven to be a dynamical battleground for competing ideas. In this arena, where some of the greatest scientific minds (and egos!) battled, we have many instances of drama, thrills, twists, and, of course, mystery:

- A priest-scientist, Fr. Georges Lemaître, interpreting his solutions as having "a day without yesterday," a progenitor term to the "Big Bang";
- One scientist's mockery of the opposing camp's view immortalized (the term "Big Bang" was coined by an ardent proponent of the steady-state theory, Sir Fred Hoyle);
- A "fudge factor" introduced, then discarded in embarrassment, then later reintroduced as a way to get the cosmic books to balance (Einstein's cosmological constant);
- The greatest experimental evidence for the Big Bang coming about by sheer accident (cosmic microwave background radiation);
- Finally, we are still searching for what comprises about 96% of the content of the Universe. Over 70% of the mass/energy content of the Universe is in the form of the unknown energy called "dark energy." Over 80% of the mass is in the form of the mysterious "dark matter."

Relativity and Cosmology
https://doi.org/10.1016/B978-0-44-323542-9.00009-2

1.1. Motivation: Newton vs. Einstein

Newtonian mechanics is an approximation that works quite well for most of our "earthly" needs, when the velocity $v \ll c$ (c is the speed of light) and when the masses are not too large. The basic differences and analogies between Newtonian and Einsteinian physics are presented in Table 1.1.

Table 1.1 Differences and analogies between Newtonian and Einsteinian mechanics.

Newton	Einstein
Absolute time and absolute space (simultaneity)	Spacetime (no simultaneity)
Galilean invariance of space	Lorentz invariance of spacetime (time dilation, length contraction)
Existence of preferred inertial frames (at rest or moving with constant velocity with respect to the absolute space)	No preferred frames (physics is the same everywhere)
Infinite speed of light c (instantaneous action at the distance)	Finite and fixed speed of light c (nothing propagates faster than c)
Gravity is a force	Gravity is a distortion of the fabric of spacetime
Newton's Second Law	Geodesic equation
Poisson equation	Einstein's equations

Newtonian mechanics quickly runs into problems that cannot be explained within its realm:

- All observers measure the same speed of light c (in a vacuum), as demonstrated by the Michelson–Morley experiment.
- Electromagnetism does not respect Galilean invariance.
- Motion near a large mass (strong gravitational field regime), such as correctly quantifying the precession of Mercury's perihelion.
- Why do all bodies experience the same acceleration regardless of their mass, *i.e.*, why is the inertial and gravitational mass the same (as measured experimentally)?

Einstein's theory of *special relativity* introduced some revolutionary concepts:

- "Abolished" absolute time – introduced four-dimensional (4D) *spacetime* as an inseparable entity.
- Set in its foundation a finite and fixed speed of light c.
- Established equivalence between energy and mass.

In the absence of gravitational effects, spacetime is flat: particles feeling no forces move along straight lines at a constant speed. 4D spacetime used in special relativity is flat, just like three-dimensional (3D) Euclidean space. Therefore special relativity is *not* a theory of gravity. Special relativity introduces a new way to do mechanics in a 4D spacetime.

Einstein's theory of *general relativity* continued the revolution:

- *Equivalence principle*: Posited the equivalence between inertial and gravitational mass – the motion in a curved spacetime is indistinguishable from the motion of an accelerated observer in flat spacetime.
- *Cosmological principle*: Our position is "as mundane as it can be" (on large spatial scales, the Universe is *homogeneous* and *isotropic*).
- *Relativity*: The laws of physics are the same everywhere.
- Massless photons are subject to gravity.

1.2. General relativity: a new theory of gravity

General relativity put forward a new definition and interpretation of gravity:

- *Gravity is a distortion of the structure of spacetime as caused by the presence of matter and energy*. The paths followed by matter and energy in spacetime are governed by the structure of spacetime. This great feedback loop is described by Einstein's field equations. John Archibald Wheeler put it eloquently: "Mass tells spacetime how to curve, and spacetime tells mass how to move." The 4D spacetime considered in general relativity is not restricted to being flat. "General" in general relativity represents a generalization to a curved spacetime from the "special" case of flat spacetime considered in special relativity.
- *Gravity is not a force!* Rather, the phenomenon that we normally call gravity is a manifestation of the fact that, in the presence of matter and energy, *spacetime is curved*.
- All sources of matter and energy feel the effects of spacetime curvature, so that (*massless!*) light experiences gravitational effects *just as matter does*. By contrast, in Newtonian mechanics, gravity is a force that affects massive objects only.[a]
- A particle that experiences no non-gravitational forces will follow a *geodesic* within the spacetime that it inhabits. For ordinary, possibly curved spaces, a geodesic is defined as the shortest distance between two points. In flat space, a geodesic is a straight line; on the surface of a sphere, it is a so-called

great circle (a circle that divides the sphere into two equal hemispheres). Most generally, equations of general relativity describe the path in which a particle moves inside an arbitrarily curved spacetime.

[a] Technically, if one assigns an equivalent "mass" to light through $E = mc^2$, the bending of light is also present in Newtonian gravity. This approach, however, is incorrect – it misses half of the effect.

After establishing general relativity as the way to describe the Universe and learning its mathematical formalism, we will finally embark on a journey of expressing mathematically the world around us on the largest scales, physically interpreting the implications and reconciling them with the observations.

Many of the phenomena for which we now have overwhelming evidence – the Big Bang, an expanding Universe, CMB radiation, black holes, among others – were *first* predicted by the solutions of Einstein's equations. Therefore it is the mathematics that holds the keys to unlocking the mysteries of the Universe, so we begin by acquiring the required mathematical skills!

1.3. Book outline

This book is composed of two parts:

1. **General relativity as the foundation of cosmology**
   Overview of the basic concepts of the theory of general relativity and the formalism it provides for studying the evolution of the Universe:
   - We begin with the foundations of *special* relativity (Chapter 2) as the necessary bridge from Newtonian gravity to Einstein's general relativity as the new theory of gravity. Special relativity leads to some strange consequences, which we will discuss in detail, such as loss of simultaneity, time dilation, and length contraction. It also establishes that mass and energy are equivalent: $E = mc^2$!
   - The Equivalence Principle (Chapter 3) establishes that gravitational effects are equivalent to acceleration. It makes a clear, testable prediction that *gravity bends light.*
   - General relativity (Chapter 4) uses tools of differential geometry: metrics, covariant and contravariant tensors, and invariants. When the equations of motion are written in a tensor form, they are invariant under metric transformation.
   - The geodesic equation determines the shortest path between two points in an arbitrary space (time). These are the paths followed by free particles.
   - Einstein's field equations (Chapter 5) quantify how the presence of matter/energy curves spacetime.
2. **Interpreting the Universe through General Relativity**
   Implications of the solutions to Einstein's equations:

- Local solutions: Schwarzschild's solution near a massive object – *black holes* (Chapter 6).
- Global solutions for the Friedmann–Lemaître–Robertson–Walker (expanding) Universe (Chapter 7):
  - Matter- and radiation-dominated Universe (Chapter 8).
  - Dark energy-dominated Universe (Chapter 9).
- Dark matter: possible candidates and the current searches (Chapter 12).
- Brief history of the early Universe, first 400,000 years, from the Big Bang to recombination (Chapter 13).
- Precision cosmology: cosmic microwave background (CMB) radiation (Chapter 14).

CHAPTER 2

Special relativity

The effort to understand the Universe is one of the very few things that lifts human life a little above the level of farce, and gives it some of the grace of tragedy.

Steven Weinberg

The Big Picture: In this chapter, we introduce a novel way to do mechanics. In Newtonian (classical) mechanics, time was absolute and separate from 3D space, thereby allowing us to express particle coordinates in terms of absolute time: $x(t)$, $y(t)$, $z(t)$. Einstein's (relativistic) mechanics – special relativity – rests on radically different foundations: 4D spacetime in which space and time form an inseparable entity. This chapter introduces the foundations of special relativity and the consequences that arise from it.

Notation:

- 3D Euclidean space *three-vectors* (x, y, z) are represented in bold italics, as in $\boldsymbol{r}$, $\boldsymbol{A}$, etc. They are also denoted as x^i, where $i = 1, 2, 3$ and $x^1 = x$, $x^2 = y$, $x^3 = z$. All indices used in three-vectors are Roman letters and go from 1 to 3.
- 4D spacetime *four-vectors* (ct, x, y, z) are shown in bold Roman letters, as in $\mathbf{q}$, $\mathbf{A}$, etc. They are also denoted as x^α, where $\alpha = 0, 1, 2, 3$ and $x^0 = ct$, $x^1 = x$, $x^2 = y$, $x^3 = z$. All indices used in four-vectors are Greek letters and go from 0 to 3.

2.1. Newtonian (non-relativistic) physics

Newtonian mechanics implements two basic concepts [1]:

1. Our ordinary, "intuitive" notion of immutable space and absolute time; and
2. The idea that only relative motion is fundamental; in particular, it is assumed that there exists a preferred set of observers, *inertial observers*, moving relative to one another with constant velocity, who see the same physical laws.

2.1.1 Inertial frames

An *inertial frame* is a frame of reference that is moving at a constant velocity. Any other frame that moves with constant velocity with respect to an inertial frame is also an iner-

Relativity and Cosmology
https://doi.org/10.1016/B978-0-44-323542-9.00010-9

tial frame itself. The two inertial frames, S and S', are said to be in *standard configuration* if the inertial frame S', with coordinates (x', y', z'), is moving at the constant velocity v along the x-axis with respect to the inertial frame S, with coordinates (x, y, z). This is shown in Fig. 2.1.

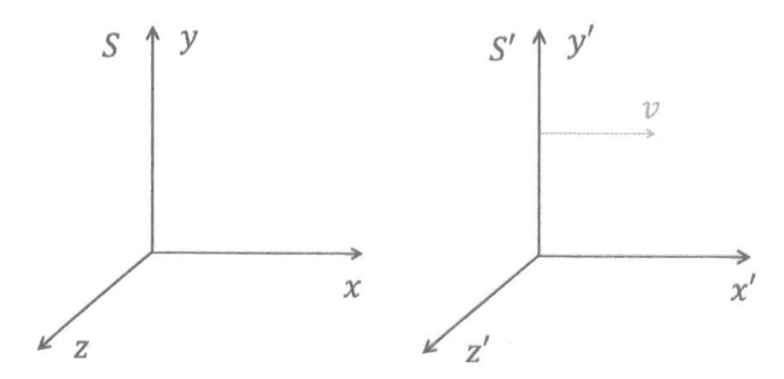

Figure 2.1 Two inertial frames in standard configuration. Primed inertial frame S' is moving at a velocity v relative to the unprimed inertial frame S along the x-axis. In Newtonian physics, time is separate from space. In Einstein's special relativity, time and space are always considered together as spacetime.

2.1.2 Galilean transformations

The geometry of space is uniquely defined by the assumption that the distance (or *line element*) dr between two nearby points $\boldsymbol{r}_1 = (x, y, z)$ and $\boldsymbol{r}_2 = (x + dx, y + dy, z + dz)$ satisfies

$$dr^2 = dx^2 + dy^2 + dz^2. \tag{2.1}$$

In particular, all of Euclidean geometry, including such ideas that (i) initially parallel lines remain parallel forever and (ii) the angles of a triangle sum up to 180°, can be derived from this relation.

This notion of invariance of distance leads to *Galilean transformations*:

1. *Translational invariance*: dr is invariant with respect to a transformation of the type

$$x' = x + k, \qquad y' = y, \qquad z' = z, \tag{2.2}$$

since $dx' = d(x + k) = dx$ (translations of y- and z-coordinates have the same form);

2. *Rotational invariance*: dr is invariant with respect to a transformation of the type

$$x' = x\cos\theta - y\sin\theta, \qquad y' = -x\sin\theta + y\cos\theta, \qquad z' = z, \tag{2.3}$$

which follows because the relations

$$dx' = dx\cos\theta - dy\sin\theta, \qquad dy' = -dx\sin\theta + dy\cos\theta, \tag{2.4}$$

imply that

$$ds^2 = dx^2 + dy^2 = dx'^2 + dy'^2 = ds'^2. \tag{2.5}$$

Again, rotations with respect to y- and z-coordinates have the same form.

Mathematically, the set of six transformations defined by these three translations and three rotations are called the *Galilean transformations*.

One also has an additional invariance with respect to time translations, *i.e.*, $t' = t + k$, but this invariance is treated on a different footing from the spatial invariances, since space and time are viewed as distinct entities.

Newtonian mechanics assumes that interactions between objects involve an *action at a distance* – objects can affect one another even though they do not touch physically. It is also assumed implicitly that this action at a distance involves *instantaneous propagation*, *i.e.*, the information about an object at $\boldsymbol{r}_1$ propagates instantaneously to another object at $\boldsymbol{r}_2$.

Laplace made an important step forward by introducing the notion of a gravitational *field*, in which interaction between objects propagates not instantaneously but at the speed of light. The interaction is carried out by waves within the field. This notion of a field is essential in our treatment of not only gravity but also electricity, magnetism, and other interactions.

It is assumed that the fundamental laws of physics follow from an Action Principle, *i.e.*, that the equations of motion can be derived by demanding that the value of some quantity S be extremal. The action can be written in terms of a Lagrangian, denoted as L or $\mathcal{L}$ and defined below, as

$$S = \int dt L(\boldsymbol{r}, \dot{\boldsymbol{r}}, t) \tag{2.6}$$

for a particle or

$$S = \int dt \int d^3r \mathcal{L}(\boldsymbol{r}, \dot{\boldsymbol{r}}, t) \tag{2.7}$$

for a field. For the case of a particle, this leads to the Lagrange equation,

$$\frac{d}{dt}\frac{\partial L}{\partial \dot{\boldsymbol{r}}} = \frac{\partial L}{\partial \boldsymbol{r}}. \tag{2.8}$$

For the simple case of a *free particle*, a particle acted on by no net force:

$$\frac{d}{dt}(m\dot{\mathbf{r}}) = 0, \tag{2.9}$$

the Lagrangian is simply

$$L = \frac{1}{2} m \dot{\mathbf{r}}^2. \tag{2.10}$$

2.1.3 Crisis of Newtonian physics

In the 19th century, three important problems arose that Newtonian mechanics could not describe:

1. The Michelson–Morley experiment: (in a vacuum) all observers measure the same speed of light c. This contradicts Galilean invariance, which predicts that, for two observers moving with relative velocity v in the x-direction,

$$x = x' + vt, \qquad y = y', \qquad z = z', \qquad t' = t, \tag{2.11}$$

so that

$$\frac{dx}{dt} = \frac{dx'}{dt} + v \quad \rightarrow \quad c = c' + v. \tag{2.12}$$

2. Electromagnetism is not invariant with respect to Galilean transformations!
   - Particles interact via retarded potentials (so-called Liénard–Wiechert potentials).
   - Galilean relativity is inconsistent with the propagation of electromagnetic radiation.
3. Precession of the perihelion of Mercury, as recognized by Urbain Le Verrier in 1859 [2].

Special relativity is an alternative to Newtonian mechanics that still assumes that the relative motion is fundamental *but* incorporates explicitly the assumptions, verified experimentally, that all observers measure the same speed of light.

2.2. Foundational premises of special relativity

There are three foundational premises of special relativity [3]:

1. *The principle of relativity.* The laws of physics are the same in all inertial frames.
2. *The speed of light is constant.* All observers observe the same speed of light, regardless of their motion.

3. *Uniform motion is invariant.* Constant motion (including being at rest as its special case) in one inertial frame transforms also into constant motion in all other inertial frames.

2.3. Lorentz transformations

Special relativity is based on the idea that the fundamental arena of physics is a 4D spacetime, characterized by an invariant *4D spacetime interval*

$$ds^2 \equiv -c^2 dt^2 + dx^2 + dy^2 + dz^2, \tag{2.13}$$

between two nearby points in 4D spacetime $\mathbf{r_1} = (ct, x, y, z)$ and $\mathbf{r_2} = (ct + cdt, x + dx, y + dy, z + dz)$.

This notion of distance can be used to derive a theory of geometry every bit as systematic as the Euclidean geometry predicated on $dr^2 = dx^2 + dy^2 + dz^2$. It is often called *Minkowskian geometry* [4], and the spacetime equipped with this notion of distance is called *Minkowski space.*

This spacetime admits 10 different symmetries, known as *Lorentz transformations* [5]:

1. *Translational invariance* in the x, y, z, and t directions (four symmetries);
2. *Space-rotational invariance*: invariance with respect to rotations in $x - y$, $x - z$, and $y - z$ planes (three symmetries);
3. *Spacetime-rotational invariance (boosts)*: invariance with respect to rotations in $x - t$, $y - t$, and $z - t$ planes (three symmetries). These transformations of the type:

$$ct' = -x \sinh\psi + ct\cosh\psi, \quad x' = x\cosh\psi - ct\sinh\psi, \quad y' = y, \quad z' = z, \tag{2.14}$$

leave the interval ds invariant. The parameter ψ is called the *rapidity*.

2.3.1 Standard configuration

Two frames S and S' in standard configuration (as shown in Fig. 2.1) are initially coincident, $x' = 0$ and $x = vt$. This condition implies

$$x' = 0 = x\cosh\psi - ct\sinh\psi = vt\cosh\psi - ct\sinh\psi = ct\left((v/c)\cosh\psi - \sinh\psi\right) \tag{2.15}$$

$$\rightarrow \qquad (v/c)\cosh\psi = \sinh\psi \qquad \rightarrow \qquad \frac{\sinh\psi}{\cosh\psi} = \tanh\psi = v/c.$$

The relation $\cosh^2\psi - \sinh^2\psi = 1$ then implies that

$$\cosh\psi = \frac{\cosh\psi}{\sqrt{1}} = \frac{\cosh\psi}{\sqrt{\cosh^2\psi - \sinh^2\psi}} = \frac{1}{\sqrt{1-\tanh^2\psi}} = \frac{1}{\sqrt{1-v^2/c^2}}, \tag{2.16}$$

$$\sinh\psi = \tanh\psi\cosh\psi = \frac{\tanh\psi}{\sqrt{1-\tanh^2\psi}} = \frac{v/c}{\sqrt{1-v^2/c^2}}, \tag{2.17}$$

so that the transform given in Eq. (2.14) becomes

$$ct' = \frac{ct-(v/c)x}{\sqrt{1-v^2/c^2}}, \qquad x' = \frac{x-(v/c)ct}{\sqrt{1-v^2/c^2}}, \qquad y' = y, \qquad z' = z. \tag{2.18}$$

Defining the *relativistic parameters* β and γ,

$$\beta \equiv \frac{v}{c}, \qquad \gamma \equiv \frac{1}{\sqrt{1-v^2/c^2}} = \frac{1}{\sqrt{1-\beta^2}}, \tag{2.19}$$

we can express the transformation above as

$$ct' = \gamma\left(ct - \beta x\right), \qquad x' = \gamma\left(x - \beta ct\right), \qquad y' = y, \qquad z' = z. \tag{2.20}$$

This is an example of a Lorentz transformation.

Lorentz transformations (just like Galilean transformations) are linear, and can be written in a compact linear algebraic format

$$\mathbf{q}' = \mathrm{L}\mathbf{q}, \tag{2.21}$$

where $\mathbf{q} \equiv (ct, x, y, z)^{\mathrm{T}}$ and $\mathbf{q}' \equiv (ct', x', y', z')^{\mathrm{T}}$ and

$$\mathrm{L} = \begin{pmatrix} \gamma & -\gamma\beta & 0 & 0 \\ -\gamma\beta & \gamma & 0 & 0 \\ 0 & 0 & 1 & 0 \\ 0 & 0 & 0 & 1 \end{pmatrix}, \tag{2.22}$$

for a Lorentz transform from the inertial frame S to the inertial frame S' in the standard configuration shown in Fig. 2.1. Transforming back from the inertial frame S' to S is given by L^{-1}:

$$\mathrm{L}^{-1} = \begin{pmatrix} \gamma & \gamma\beta & 0 & 0 \\ \gamma\beta & \gamma & 0 & 0 \\ 0 & 0 & 1 & 0 \\ 0 & 0 & 0 & 1 \end{pmatrix}. \tag{2.23}$$

Here, the superscript T denotes a linear transpose, making $(, , ,)^{\mathrm{T}}$ a column vector.

Note that for low velocities, $v \ll c$, so $\beta \approx 0$ and $\gamma \approx 1$, one recovers an approximately Galilean transformation, for which

$$t' \approx t, \qquad x' \approx x - vt, \qquad y' = y, \qquad z' = z. \tag{2.24}$$

This is important because of the overwhelming experimental evidence that Newtonian physics is almost exactly correct for $v \ll c$.

The magnitude of the relativistic parameter $\beta = v/c$ can be viewed as a criterion for the need for relativistic treatment: if it is not negligible, Newtonian mechanics is insufficient, and fully relativistic treatment is necessary.

2.3.2 Composition of velocities and general Lorentz transformations

Any two inertial frames are connected by a linear transform of the form given in Eq. (2.21). Owing to their linearity, both Lorentz and Galilean transformations can be composed to form another transformation: if the inertial frame S_2 is related to the inertial frame S_1 via $\mathbf{q}_2 = \mathrm{L}_{12}\mathbf{q}_1$, and the inertial frame S_3 is related to the inertial frame S_2 by $\mathbf{q}_3 = \mathrm{L}_{23}\mathbf{q}_2$, then the relationship between inertial frames S_3 and S_1 is given by $\mathbf{q}_3 = \mathrm{L}_{23}\mathrm{L}_{12}\mathbf{q}_1$.

A general Lorentz transform between two inertial frames moving with respect to each other with a constant velocity $\boldsymbol{v} = v\boldsymbol{n}$ is given by:

$$ct' = \gamma\left(ct - \beta \boldsymbol{n} \cdot \boldsymbol{q}\right), \tag{2.25}$$

$$\boldsymbol{q}' = \boldsymbol{q} + (\gamma - 1)(\boldsymbol{q} \cdot \boldsymbol{n})\boldsymbol{n} - \gamma\beta ct\boldsymbol{n}, \tag{2.26}$$

where $\boldsymbol{q} = (x, y, z)$, $\boldsymbol{q}' = (x', y', z')$, $\boldsymbol{n} = \frac{1}{\beta}(\beta_x, \beta_y, \beta_z)$, $\beta_x = v_x/v$, $\beta_y = v_y/v$, $\beta_z = v_z/v$, $v^2 = v_x^2 + v_y^2 + v_z^2$ and $\beta^2 = \beta_x^2 + \beta_y^2 + \beta_z^2$.

The Lorentz transformations imply phenomena like *length contraction* and *time dilation.*

2.4. Length contraction

Suppose that there is a rod at rest in the inertial frame S', moving at v along the x-direction with respect to the inertial frame S. To determine the length of the rod, an observer in the inertial frame S must determine the location of the two ends of the rod x_1 and x_2 at the same instant t. This means that

$$x_1' = \gamma\left(x_1 - \beta ct\right), \qquad x_2' = \gamma\left(x_2 - \beta ct\right) \quad \rightarrow \quad \Delta x' = \gamma\Delta x. \tag{2.27}$$

If ℓ_0 is the *proper length* of the rod – the length as measured in a frame of reference in which the rod is at rest, which in this case is S', so $\ell_0 \equiv \Delta x'$ – then the length $\ell(v) \equiv \Delta x$ as measured in the moving frame will satisfy

$$\ell(v) = \frac{\ell_0}{\gamma} = \ell_0\sqrt{1-\beta^2} = \ell_0\sqrt{1-v^2/c^2}. \tag{2.28}$$

This means that in the reference frame in which the rod appears to move, the rod appears shorter! It is shortened along the direction of motion of the moving frame. Note also that, because there is no contraction in the two spatial directions orthogonal to the motion, the spatial volume in the moving frame $V(v)$ appears to contract from its volume in the rest frame V_0 by an identical amount: $V(v) = V_0\sqrt{1-v^2/c^2}$.

2.5. Time dilation

In the similar fashion, to determine the duration of some process, one must measure the initial and final times at some fixed point in space x'. An interval of time as measured by an observer in the inertial frame S is

$$ct_1 = \gamma\left(ct_1' + \beta x'\right), \qquad ct_2 = \gamma\left(ct_2' + \beta x'\right) \quad \rightarrow \quad \Delta t = \gamma \Delta t'. \tag{2.29}$$

This means that in the reference frame in which the rod appears to move, S, the time is longer – time intervals are dilated: $\Delta t > \Delta t'$. Eq. (2.29) has been derived assuming that v is constant, but that was by no means necessary. The entire calculation can be reformulated infinitesimally, in which case, by integrating over a finite interval, one arrives at

$$t_2' - t_1' = \int_{t_1}^{t_2} \sqrt{1 - v^2(t)/c^2}\, dt. \tag{2.30}$$

2.6. Invariance of the volume element

Do observers in different inertial frames measure a different volume element $d\Omega = dxdydzcdt$?

Again, consider two inertial frames, S and S', in standard configuration, where the inertial frame S' moves at v along the x-direction with respect to the inertial frame S. From the previous two sections we saw that the measurements in the rest frame S will experience length contraction (Eq. (2.27)),

$$dx = \frac{dx'}{\gamma}, \tag{2.31}$$

and time dilation (Eq. (2.29)),

$$dt = \gamma\, dt'. \tag{2.32}$$

The measurements in the other two remaining directions, normal to the direction of motion (y-, z-directions) are not affected:

$$dy = dy', \qquad dz = dz'. \tag{2.33}$$

The volume element in the inertial frame S is, therefore,

$$dxdydzcdt = \frac{dx'}{\gamma} dy' dz' \gamma\, dt' = dx' dy' dz' cdt', \tag{2.34}$$

the same as the volume element measured in the inertial frame S'. Therefore *all inertial observers agree on the magnitude of the 4D volume element* $d\Omega = dxdydzcdt$.

2.7. Timelike, spacelike, and null intervals

Note that the 4D spacetime interval

$$ds^2 = -c^2 d\tau^2 = -c^2 dt^2 + dx^2 + dy^2 + dz^2 \tag{2.35}$$

is not of uniform sign! Here, $d\tau$ is the *proper time* – time measured in a system at rest. One can rewrite the equation above as

$$ds^2 = -c^2 d\tau^2 = -c^2 dt^2 \left(1 - \frac{dx^2 + dy^2 + dz^2}{c^2 dt^2}\right) = -c^2 dt^2 \left(1 - \frac{v^2}{c^2}\right). \tag{2.36}$$

It therefore follows that:

- If $v^2 < c^2$, then $ds^2 < 0$, which is called a *timelike interval*;
- If $v^2 > c^2$, then $ds^2 > 0$, which is called a *spacelike interval*;
- If $v^2 = c^2$, then $ds^2 = 0$, which is called a *null interval*, *i.e.*, "light never goes anywhere."

All objects with non-zero mass must follow trajectories corresponding to timelike intervals, for which the locally measured speed $v < c$. Massless photons follow null intervals, for which $ds = 0$, and, therefore, the proper time $d\tau = 0$: light does not "experience" time!

For a timelike interval, $v < c$, one can write

$$c^2 d\tau^2 = c^2 dt^2 \left(1 - v^2/c^2\right) \quad \rightarrow \quad d\tau = dt\sqrt{1 - v^2/c^2} = dt/\gamma. \tag{2.37}$$

The proper time $d\tau$, measured by the observer at rest, is always the shortest. Therefore we have recovered time dilation.

2.8. Recovering Newtonian mechanics: special limit $v \ll c$

The motion of a free particle derives from an action

$$S = -mc^2 \int d\tau = -mc^2 \int \frac{d\tau}{dt} dt = -mc^2 \int \sqrt{1 - v^2/c^2}\, dt \tag{2.38}$$
$$\rightarrow \quad L = -mc^2 \sqrt{1 - v^2/c^2} \approx -mc^2 + \frac{1}{2} mv^2,$$

i.e., the equation of motion derives from the demand that the proper time/length be extremized. In the last step, we used the binomial expansion for speeds much smaller than the speed of light (small values of v/c) to approximate

$$\sqrt{1 - v^2/c^2} = \left(1 - v^2/c^2\right)^{1/2} \approx 1 - \frac{1}{2} v^2/c^2. \tag{2.39}$$

The canonical momentum $\boldsymbol{p}$ satisfies

$$\boldsymbol{p} \equiv \frac{\partial L}{\partial \boldsymbol{v}} = \frac{m\boldsymbol{v}}{\sqrt{1 - v^2/c^2}} \quad \rightarrow \quad \frac{d\boldsymbol{p}}{dt} = 0 = \frac{d}{dt} \frac{m\boldsymbol{v}}{\sqrt{1 - v^2/c^2}}. \tag{2.40}$$

The Hamiltonian function, which is equal numerically to the energy E, satisfies

$$H = \boldsymbol{p} \cdot \boldsymbol{v} - L = \sqrt{m^2c^4 + p^2c^2} \approx mc^2 + \frac{p}{2mc} \quad \text{for} \quad v \ll c. \tag{2.41}$$

Hence, a massive object, as $v \rightarrow c$, has momentum $p \rightarrow \infty$, and, therefore, has energy $E \rightarrow \infty$!

The *rest mass*, m_0, of a particle is the mass of the particle as measured in the inertial frame in which the particle is at rest. If a particle is moving with respect to an observer in an inertial frame S with velocity v, then the observer measures the mass of the particle to be

$$m = \frac{m_0}{\sqrt{1 - v^2/c^2}} = \gamma m_0. \tag{2.42}$$

Again, for $v \ll c$, *i.e.*, when $v/c \rightarrow 0$, binomial expansion can be used to approximate

$$\frac{1}{\sqrt{1 - v^2/c^2}} = \left(1 - v^2/c^2\right)^{-1/2} \approx 1 + \frac{1}{2} v^2/c^2, \tag{2.43}$$

so that the mass of a particle moving with the respect to the observer is measured by that observer to be

$$m \approx m_0 \left(1 + \frac{1}{2} v^2/c^2\right) = m_0 + \frac{1}{2} m_0 v^2/c^2, \tag{2.44}$$

or, multiplying by c^2,

$$mc^2 \approx m_0 c^2 + \frac{1}{2} m_0 v^2 = E. \tag{2.45}$$

Therefore we recover the result of Newtonian mechanics: the total energy of a particle is equal to the sum of its rest energy and its Newtonian kinetic energy.

2.9. Covariant and contravariant four-vectors

Why does one care about vectors? When written in vectorial form, the equations of motion formulated in the context of the Newtonian physics are invariant with respect to Galilean transformations. In a similar fashion, the laws of special relativity can be written in terms of suitably defined vectors in such a fashion that they are invariant under Lorentz transformations.

In Minkowski space, a *four-vector* is a collection of four numbers that transforms like the spatial coordinates $x^\alpha = (x^0, x^1, x^2, x^3)^\mathrm{T} = (ct, x, y, z)^\mathrm{T}$ under a Lorentz transformation. Therefore an arbitrary vector $A^\alpha = (A^0, A^1, A^2, A^3)^\mathrm{T}$ transforms according to Eq. (2.20). Given this definition, one can interpret

$$ds^2 = -c^2 dt^2 + dx^2 + dy^2 + dz^2, \tag{2.46}$$

as an equation involving four-vectors, writing

$$ds^2 = \sum_{\alpha=0}^{3} \sum_{\beta=0}^{3} g_{\alpha\beta} \, dx^\alpha \, dx^\beta, \tag{2.47}$$

where

$$g_{\alpha\beta} = \begin{pmatrix} -1 & 0 & 0 & 0 \\ 0 & 1 & 0 & 0 \\ 0 & 0 & 1 & 0 \\ 0 & 0 & 0 & 1 \end{pmatrix}. \tag{2.48}$$

$g_{\alpha\beta}$ is called the *spacetime metric*, since it serves to define distances between nearby points.[1]

[1] Another *metric signature* – non-zero elements of the Minkowski spacetime metric – also exists: $(1, -1, -1, -1)$.

It is conventional to implement the so-called *Einstein's summation convention*, where there is an implicit summation when the same index appears twice, *i.e.*, so that one could rewrite Eq. (2.47) simply as

$$ds^2 = g_{\alpha\beta} dx^\alpha dx^\beta. \tag{2.49}$$

Technically, dx^α is termed a *contravariant vector* or a *contravariant first-rank tensor.* $g_{\alpha\beta}$ is termed a *covariant second-rank tensor.* (Contravariant means the indices α, β, *etc.* "live" upstairs; covariant means that the indices "live" downstairs. First rank means that there is one index; second rank means there are two indices.) Pragmatically, the quantity dx^α can be viewed as a column matrix (column vector) and $g_{\alpha\beta}$ can be viewed as a square matrix. More formally, dx^α and $g_{\alpha\beta}$ are examples of so-called *tensors.*

The quantity $dx_\beta \equiv dx^\alpha g_{\alpha\beta}$ can be viewed as a row matrix (row vector). It is a simple example of a *covariant vector* or a *covariant first rank tensor.* Note that if dx^α has components ct, x, y, and z, then dx_α has components $-ct$, x, y, and z.

A *covariant vector* is a collection of four numbers that transform like $dx_\alpha = g_{\alpha\beta} dx^\beta$ under Lorentz transformations. Therefore an arbitrary covariant vector $A_\alpha = (A_0, A_1, A_2, A_3)$ transforms as

$$A_0 = \gamma\left(A'_0 - \beta A'_1\right), \qquad A_1 = \gamma\left(A'_1 - \beta A'_0\right), \qquad A_2 = A'_2, \qquad A_3 = A'_3. \tag{2.50}$$

A *contravariant vector* is a collection of four numbers that transform like $dx^\alpha = g^{\alpha\beta} dx_\beta$ under Lorentz transformations. An arbitrary contravariant vector $A^\alpha = (A^0, A^1, A^2, A^3)^{\mathrm{T}}$ transforms as

$$A^0 = \gamma\left(A'^0 + \beta A'^1\right), \qquad A^1 = \gamma\left(A'^1 + \beta A'^0\right), \qquad A^2 = A'^2, \qquad A^3 = A'^3. \tag{2.51}$$

The length of dx^α satisfies $ds^2 = g_{\alpha\beta} dx^\alpha dx^\beta$, so, by analogy, for a general vector A^α,

$$|A|^2 \equiv g_{\alpha\beta} A^\alpha A^\beta = A_\beta A^\beta = -\left(A^0\right)^2 + \left(A^1\right)^2 + \left(A^2\right)^2 + \left(A^3\right)^2. \tag{2.52}$$

Analogously, one can define a *scalar product*

$$\mathbf{A} \cdot \mathbf{B} \equiv g_{\alpha\beta} A^\alpha B^\beta = A_\beta B^\beta = A^\alpha B_\alpha. \tag{2.53}$$

The scalar product in Eq. (2.53) is a 4D spacetime analogue of the dot product in ordinary 3D Euclidean space, *i.e.*,

$$\boldsymbol{A} \cdot \boldsymbol{B} \equiv \delta_{ij} A^i B^j = A_j B^j = A^i B_i, \tag{2.54}$$

where δ_{ij} is the usual Euclidean flat-space metric:

$$\delta_{ij} = \begin{pmatrix} 1 & 0 & 0 \\ 0 & 1 & 0 \\ 0 & 0 & 1 \end{pmatrix}. \tag{2.55}$$

Note that $\mathbf{A} \cdot \mathbf{B}$ is invariant, *i.e.*, "transforms as a scalar" under Lorentz transformations.

2.9.1 Four-velocity

The ordinary three-velocity $\boldsymbol{v} = d\mathbf{x}/dt$, with components $v^i = dx^i/dt$, as defined in Newtonian mechanics, does not transform as a vector in Minkowski space and, as such, is *not* a natural object to consider in special relativity. Instead, it is natural to define a *four-velocity*,

$$u^\alpha = \frac{dx^\alpha}{d\tau}. \tag{2.56}$$

As dx^α transforms as a vector under Lorentz transformations and the proper time $d\tau$ is invariant under Lorentz transformations, u^α transforms as a vector. u^α is easily related to the ordinary three-velocity in a specific frame of reference.

> *One of the fundamental assumptions of special relativity is that rates of all moving clocks depend only on their velocity, so that an accelerated clock will be time dilated in the same way as a uniformly moving one with the same instantaneous speed.*

Thus using $dt = d\tau/\sqrt{1 - v^2/c^2}$, one has

$$u^0 = \frac{dct}{d\tau} = c\frac{dt}{d\tau} = c\,\frac{1}{\sqrt{1 - v^2/c^2}} = \gamma c \tag{2.57}$$

and

$$u^1 = \frac{dx}{d\tau} = \frac{dx}{dt}\frac{dt}{d\tau} = \frac{v_x}{\sqrt{1 - v^2/c^2}} = \gamma v_x, \tag{2.58}$$

where, again, $\gamma = \sqrt{1 - v^2/c^2}$. The four-velocity thus has components

$$u^\alpha = (\gamma c, \gamma \boldsymbol{v})^{\mathrm{T}} = (\gamma c, \gamma v_x, \gamma v_y, \gamma v_z)^{\mathrm{T}} = \mathbf{u}. \tag{2.59}$$

Note that the scalar product, invariant under coordinate transform, of **u** with itself is

$$\begin{aligned}\mathbf{u}\cdot\mathbf{u} &= g_{\alpha\beta}u^{\alpha}u^{\beta} = -(u^0)^2 + (u^1)^2 + (u^2)^2 + (u^3)^2 \\ &= -\gamma^2c^2 + \gamma^2\left(v_x^2 + v_y^2 + v_z^2\right) = -\gamma^2c^2\left(1 - v^2/c^2\right) = -c^2.\end{aligned} \tag{2.60}$$

2.10. Energy–momentum four-vector

In the absence of forces, particles move with constant speed. This is equivalent to the statement that

$$\frac{d\mathbf{u}}{d\tau} = 0, \tag{2.61}$$

where $d\mathbf{u}/d\tau$ defines *four-acceleration.*

How should this be generalized to an analog of Newton's Second Law $\boldsymbol{F} = m\boldsymbol{a}$? Any plausible rule must satisfy three criteria:

1. It must satisfy the principle of relativity: have the same form in every inertial frame.
2. It must reduce to $d\mathbf{u}/d\tau = 0$ for a vanishing "force."
3. It must reduce to $\boldsymbol{F} = m\boldsymbol{a}$ for speeds $v \ll c$.

The obvious candidate is

$$m_0\frac{d\mathbf{u}}{d\tau} = \mathbf{F}, \tag{2.62}$$

where m_0 is the particle's *rest mass*, and **F** is the *four-force.* This clearly satisfies criteria 1 and 2. For appropriate choices of **F**, criterion 3 is also satisfied.

By analogy with the Newtonian mechanics, one is then led to identify the relativistic *energy–momentum four-vector* (or *four-momentum*),

$$\mathbf{p} = m_0\mathbf{u} = m_0\gamma(c, \boldsymbol{v})^{\mathrm{T}}, \tag{2.63}$$

which has components

$$p^0 = \frac{m_0c}{\sqrt{1 - v^2/c^2}}, \qquad \text{and} \qquad p^i = \frac{m_0v^i}{\sqrt{1 - v^2/c^2}}, \qquad i = 1, 2, 3. \tag{2.64}$$

For small velocities, $v \ll c$, the relativistic factor is

$$\gamma = \frac{1}{\sqrt{1 - v^2/c^2}} = \left(1 - v^2/c^2\right)^{-1/2} \approx 1 + \frac{1}{2}v^2/c^2, \tag{2.65}$$

so the first element of the four-momentum becomes

$$p^0 = m_0 c + \frac{1}{2} m_0 v^2/c, \qquad \text{and} \qquad \boldsymbol{p} = m_0 \boldsymbol{v}, \tag{2.66}$$

that is, $\boldsymbol{p}$ reduces to the usual Euclidean three-momentum and cp^0 yields the rest energy plus the kinetic energy $E = m_0 c^2 + (1/2) m_0 v^2$. The dot product of $\mathbf{p}$ with itself is

$$\mathbf{p} \cdot \mathbf{p} = -E^2/c^2 + p_x^2 + p_y^2 + p_z^2 = -E^2/c^2 + p^2. \tag{2.67}$$

After recalling that the dot product is a scalar, and therefore invariant under Lorentz transformation, we are free to evaluate Eq. (2.67) in any inertial frame. While special relativity holds that there are no preferred inertial frames – physics is the same in all of them – there certainly are more convenient inertial frames that minimize the number of computations necessary to obtain the same result. The most convenient inertial frame for computing invariants, such as dot products or spacetime intervals, is the rest frame of the particle in which $p_x = p_y = p_z = 0$.

Therefore the dot product of the four-momentum $\mathbf{p}$ with itself is

$$\mathbf{p} \cdot \mathbf{p} = -E^2/c^2 + p^2 = -m_0^2 c^2, \tag{2.68}$$

which leads to Einstein's famous formula for a particle's rest energy,

$$E_0 = m_0 c^2, \tag{2.69}$$

establishing an equivalence between the rest energy and the rest mass.

Combining Eqs. (2.68) and (2.69) leads to the expression for the total energy of a relativistic particle,

$$E^2 = m_0^2 c^4 + p^2 c^2. \tag{2.70}$$

Particles with $m_0 = 0$, like the photon, have both energy and momentum, and, as such, a non-vanishing energy–momentum four-vector. They thus satisfy

$$\mathbf{p} \cdot \mathbf{p} = 0. \tag{2.71}$$

They can only satisfy $p^0 = m_0/\sqrt{1 - v^2/c^2}$ and $\boldsymbol{p} = m_0 \boldsymbol{v}/\sqrt{1 - v^2/c^2}$ in the singular limit that $v \to c$, so one can infer that all massless particles travel with the speed of light c. Therefore the four-momentum for a photon with a wavelength λ, frequency

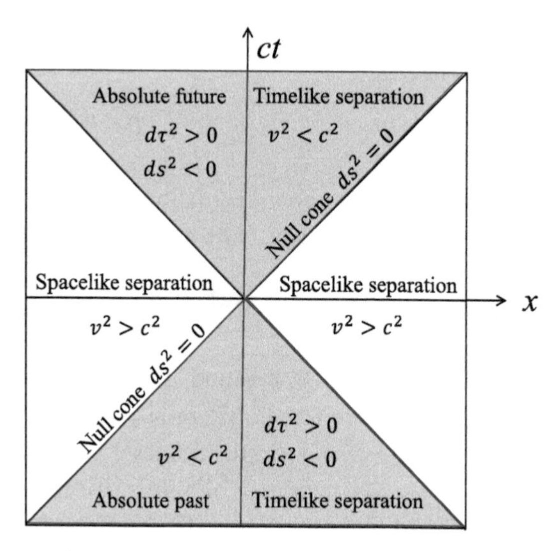

Figure 2.2 A spacetime diagram. Coordinate *x* represents the spatial, and *ct* the temporal, component of spacetime. Multiplying the time *t* by the speed of light *c* puts *ct* in the same dimensions of length as the spatial coordinates.

$\nu = c/\lambda$, and energy $E = h\nu = hc/\lambda$ (where h is the Planck constant) is

$$\mathbf{p} = \frac{E}{c}(1, \beta_x, \beta_y, \beta_z)^{\mathrm{T}} = \frac{h\nu}{c}(1, \beta_x, \beta_y, \beta_z)^{\mathrm{T}} = \frac{h}{\lambda}(1, \beta_x, \beta_y, \beta_z)^{\mathrm{T}}. \tag{2.72}$$

Here, $\beta_{x,y,z} = v_{x,y,z}/c$ and $\beta_x^2 + \beta_y^2 + \beta_z^2 = 1$, in order to satisfy the requirement that photons move at the speed of light c.

For the frames S and S' in the standard configuration (Fig. 2.1), the energy–momentum four-vector,

$$\mathbf{p} = m_0\mathbf{u} = (E/c, p_x, p_y, p_z)^{\mathrm{T}} \tag{2.73}$$

transforms as:

$$E' = \gamma E - \beta\gamma c p_x, \qquad p'_x = \gamma p_x - \beta\gamma E/c, \qquad p'_y = p_y, \qquad p'_z = p_z. \tag{2.74}$$

2.11. Spacetime diagrams

Kinematics involves specifying the trajectories of objects through spacetime, parametrized appropriately. The natural parameter is *proper time*, *i.e.*, time measured by a clock that the object carries along with it (in the object's rest frame). This gives rise to the concept of a *worldline* moving through spacetime, which corresponds to specifying four coordinates as functions of proper time τ,

$$x^\alpha = x^\alpha(\tau). \tag{2.75}$$

A useful fashion in which to visualize a particle trajectory is in the context of a *spacetime diagram*, shown in Fig. 2.2. Suppose, for simplicity, that one works in a frame of reference where the only spatial motion is in the x-direction. It is then natural to generate a plot of the $ct - x$ plane, to which the spacetime motion is necessarily restricted.

In such a plot, trajectories with $v = c$ correspond to straight lines with slope ± 1. If one were somehow to plot all three spatial dimensions, these lines would entail cuts through a hypercone, so that these trajectories of slope unity are called the *light cone*.

A particle that is at rest in the specified frame of reference has an infinite slope, and corresponds to a vertical line. Physical trajectories with $v < c$ that pass through $x = 0$ at $ct = 0$ must always travel with slope of magnitude greater than unity, *i.e.*, travel from the origin into the *absolute future* and arrive at the origin from the *absolute past*. Trajectories with slopes of magnitude less than unity are unphysical, since they would correspond to particles traveling with speeds $v > c$.

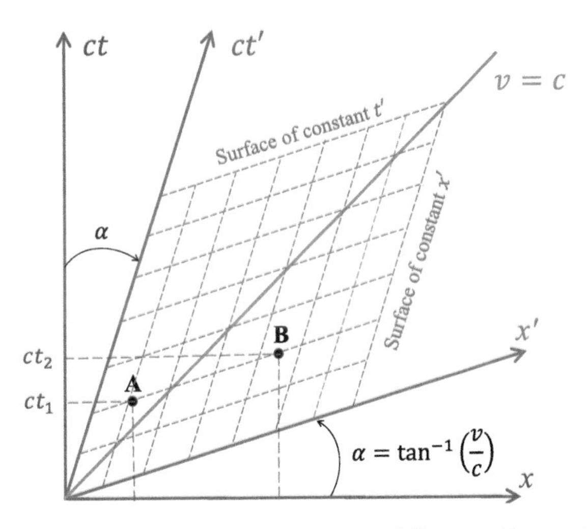

Figure 2.3 Connection between two different inertial frames. Events A and B happened simultaneously in the inertial frame (ct', x'), but in the inertial frame (ct, x), event A happened before event B ($t_1 < t_2$).

Suppose now that one considers a different frame of reference moving relative to the original frame of reference with speed v in the x-direction, but that agrees as to the location of the origin $ct = x = 0$. Spacetime coordinates in that new frame can be identified in the original spacetime diagram if one recognizes that the old and new coordinates ct, x and ct', x' are related by Lorentz transformation, which corresponds to a rotation in Minkowski spacetime. If the Lorentz transformation were a rotation in Euclidean space, all that would happen is that the new frame of reference involves a uniform rotation about the origin by some angle α. Given, however, that one is dealing with a rotation in Minkowski spacetime, things are a little more subtle. Specifically, what one finds is that, for $v > 0$, the ct'- and x'-axes have been rotated *toward one another* in such a fashion that the angle between them, as computed in the original frame of reference, is smaller than 90°, $\tan^{-1}(v/c)$ to be exact. The higher the speed v, the more compressed around the diagonal the coordinate system (ct', x') will be. The limiting case is the total collapse of the (ct', x') coordinate system for $v = c$ into a single diagonal line, the null interval. This is shown in Fig. 2.3.

Horizontal lines in a spacetime diagram correspond to *surfaces of simultaneity*. In other words, all points on these surfaces correspond in the specified frame of reference to the same instant of coordinate time t. If, in the original spacetime diagram, one exhibits surfaces of simultaneity as defined in another inertial frame, these surfaces will not, in general, correspond to horizontal lines. What this means is that different observers moving relative to one another will not agree as to whether two events happened at the same time or even which of two events happened first.

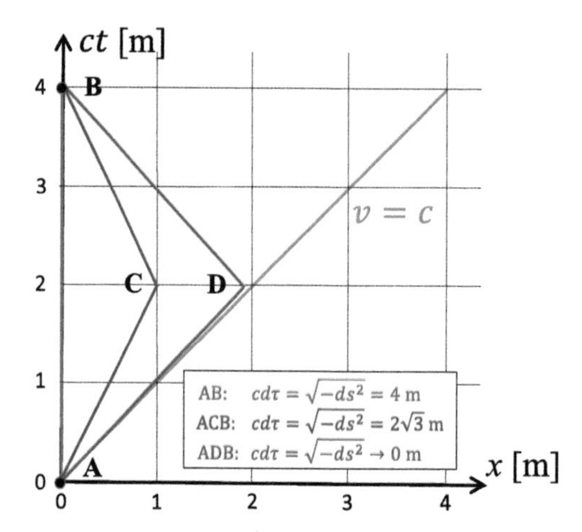

Figure 2.4 Geodesics in spacetime. The geodesic path between two points A and B is the one that *maximizes* the proper time $d\tau$ measured by a particle taking that path. Here, the path AB is a geodesic. The path ACB takes a shorter proper time. A particle on the path ADB travels nearly at the speed of light, and thereby has a vanishingly small proper time. *Photons do not experience time!*

2.12. Geodesics: equations of motion from an action principle

In spacetime, a free particle follows a worldline between two points that extremizes the spacetime interval. This path is called a *geodesic*. In 3D space, the geodesic *minimizes* the length of the space interval $dl^2 = dx^2 + dy^2 + dz^2$ between two points, thereby defining the shortest spatial distance between them. In 4D spacetime, however, the geodesic *maximizes* the proper time $d\tau$ measured by a particle taking that path. In terms of the spacetime interval, $ds^2 = -c^2 d\tau^2 = -c^2 dt^2 + dx^2 + dy^2 + dz^2$, or $d\tau = \sqrt{-ds}/c$. This is illustrated in Fig. 2.4.

Consider an arbitrary path between point A and B. The elapsed proper time along this path is

$$\tau_{\mathrm{AB}} = \int_{\mathrm{A}}^{\mathrm{B}} d\tau = \frac{1}{c}\int_{\mathrm{A}}^{\mathrm{B}} \sqrt{c^2 dt^2 - dx^2 - dy^2 - dz^2}. \tag{2.76}$$

Suppose now that instead of the proper time, the path is parametrized by another parameter λ, so chosen that it takes a fixed value, say $\lambda = 0$, at a point A and another fixed value, say $\lambda = 1$, at point B. The path is then specified by giving the coordinates as a function of λ, *i.e.*, $x^\alpha = x^\alpha(\lambda)$. One thus has

$$\begin{aligned}\tau_{\mathrm{AB}} &= \int_0^1 d\lambda \frac{d\tau}{d\lambda} = \frac{1}{c}\int_0^1 d\lambda \sqrt{\left(\frac{cdt}{d\lambda}\right)^2 - \left(\frac{dx}{d\lambda}\right)^2 - \left(\frac{dy}{d\lambda}\right)^2 - \left(\frac{dz}{d\lambda}\right)^2} \\ &= \frac{1}{c}\int_0^1 d\lambda \sqrt{-g_{\alpha\beta}\frac{dx^\alpha}{d\lambda}\frac{dx^\beta}{d\lambda}} \equiv \frac{1}{c}\int_0^1 d\lambda L.\end{aligned} \tag{2.77}$$

What one wants is to determine the path (or paths) that extremizes τ_{AB}, *i.e.*, for which a small variation $\delta x^\alpha(\lambda)$ produces a vanishing variation in elapsed proper time, *i.e.*, $\delta\tau_{\mathrm{AB}} = 0$.

This is a standard problem from classical mechanics, which leads to the *Lagrange equations* [6]:

$$\frac{d}{d\lambda}\left(\frac{\partial L}{\partial(dx^\alpha/d\lambda)}\right) - \frac{\partial L}{\partial x^\alpha} = 0 \quad \text{with } L = \sqrt{-g_{\alpha\beta}\frac{dx^\alpha}{d\lambda}\frac{dx^\beta}{d\lambda}}. \tag{2.78}$$

Solving the Lagrange equations yields

$$\frac{d}{d\lambda}\left(\left(-g_{\alpha\beta}\frac{dx^\alpha}{d\lambda}\frac{dx^\beta}{d\lambda}\right)^{-1/2} g_{\alpha\beta}\frac{dx^\beta}{d\lambda}\right) = 0. \tag{2.79}$$

However, $(-g_{\alpha\beta}(dx^{\alpha}/d\lambda)(dx^{\beta}/d\lambda))^{1/2} = cd\tau/d\lambda$, so that, multiplying by $cd\lambda/d\tau$ and lowering the index α, one has

$$\frac{cd\lambda}{d\tau}\frac{d}{d\lambda}\left(\frac{d\lambda}{cd\tau}g_{\alpha\beta}\frac{dx^{\beta}}{d\lambda}\right) = \frac{cd}{d\tau}\left(\frac{d\lambda}{cd\tau}\frac{dx_{\alpha}}{d\lambda}\right) = \frac{d^2x_{\alpha}}{d\tau^2} = 0. \tag{2.80}$$

Upon integrating twice, one finally recovers Newton's Second Law: a trajectory of a free particle (one upon which no net force acts) is a straight line $x_{\alpha} = c_1\tau + c_2$, with c_1 and c_2 some constants.

Geodesics in Minkowski spacetime are straight lines.

2.13. What an observer observes

An observer carries along four orthogonal unit vectors $\mathbf{e}_0$, $\mathbf{e}_1$, $\mathbf{e}_2$, and $\mathbf{e}_3$, which define, respectively, their time direction and three spatial directions relative to which the observer makes observations. The timelike unit vector $\mathbf{e}_0$ will be tangent to the observer's worldline, since that is the spacetime direction in which their clocks are moving. Therefore the unit vector $\mathbf{e}_0$ will be parallel to the four-velocity $\mathbf{u} = (\gamma c, \gamma v_x, \gamma v_y, \gamma v_z)^{\mathrm{T}}$. Recall from Eq. (2.60) that $\mathbf{u}\cdot\mathbf{u} = -c^2$, which means that $((1/c)\mathbf{u})\cdot((1/c)\mathbf{u}) = -1$. Therefore $\mathbf{e}_0 = (1/c)\mathbf{u} = (\gamma, v_x/c, v_y/c, v_z/c)^{\mathrm{T}}$.

All the results of all measurements of an observer can be computed in terms of the projections of physical quantities onto the orthogonal unit vectors that define the inertial frame of the observer.

Example 1: The measured energy of a particle is

$$E = -\mathbf{p}\cdot\mathbf{u}, \tag{2.81}$$

where $\mathbf{p}$ is the four-momentum of the particle being observed and $\mathbf{u}$ is the four-velocity of the observer. Note that the energy E defined in this fashion is a scalar, *i.e.*, invariant under Lorentz transformations, so that this result holds in any frame of reference. Suppose, therefore, that one considers a frame of reference in which the particle is at rest and the observer moves with velocity $\boldsymbol{v}$ in the x-direction. One then has

$$\mathbf{p} = (m_0c, 0, 0, 0)^{\mathrm{T}} \quad \text{and} \quad \mathbf{u} = (\gamma c, \gamma v, 0, 0)^{\mathrm{T}}. \tag{2.82}$$

It thus follows that the energy of the particle measured by the observer is

$$E = -\mathbf{p}\cdot\mathbf{u} = -g_{\alpha\beta}p^{\alpha}u^{\beta} = -g_{00}(m_0c)(\gamma c) = \gamma m_0c^2, \tag{2.83}$$

where we used $g_{00} = -1$ from Eq. (2.48).

Example 2: In one particular inertial frame, a particle is moving to the right with $v = c/2$. An observer whose worldline intersects the particle worldline moves with speed $v = 4c/5$. What energy does the observer measure?
Velocity $v = c/2$ corresponds to $\gamma = 1/\sqrt{1 - v^2/c^2} = \sqrt{4/3} = 2/\sqrt{3}$; velocity $v = 4c/5$ yields $\gamma = 5/3$. Thus the particle four-momentum $\mathbf{p}$ and the observer four-velocity $\mathbf{u}$ satisfy

$$\mathbf{p} = m_0 c \left(\frac{2}{\sqrt{3}}, \frac{1}{\sqrt{3}}, 0, 0 \right)^{\mathrm{T}} \quad \text{and} \quad \mathbf{u} = c \left(\frac{5}{3}, \frac{4}{3}, 0, 0 \right)^{\mathrm{T}}. \tag{2.84}$$

It follows that

$$E = -\mathbf{p} \cdot \mathbf{u} = -g_{\alpha\beta} p^{\alpha} u^{\beta} = -m_0 c^2 \left(g_{00} \frac{2}{\sqrt{3}} \frac{5}{3} + g_{11} \frac{1}{\sqrt{3}} \frac{4}{3} \right) = \frac{2}{\sqrt{3}} m_0 c^2. \tag{2.85}$$

In a similar fashion, the measured spatial momentum in the $\mathbf{e}_{\alpha}$ direction is $P_{\alpha} = \mathbf{p}\mathbf{e}_{\alpha}$.

Example 3: Show that it is impossible for an isolated free electron to admit or absorb a photon.
Conservation of four-momentum implies that

$$\mathbf{p}_{\gamma} + \mathbf{p}_{e} = \mathbf{p}_{e}', \tag{2.86}$$

where $\mathbf{p}_{\gamma}$ represents the photon four-momentum and $\mathbf{p}_{e}$ and $\mathbf{p}_{e}'$ represent the initial and final electron momenta. Squaring the left-hand side of the equality yields

$$(\mathbf{p}_{\gamma} + \mathbf{p}_{e})^2 = \mathbf{p}_{\gamma} \cdot \mathbf{p}_{\gamma} + 2\mathbf{p}_{\gamma} \cdot \mathbf{p}_{e} + \mathbf{p}_{e} \cdot \mathbf{p}_{e} = 0 + 2\mathbf{p}_{\gamma} \cdot \mathbf{p}_{e} - m_e^2 c^2, \tag{2.87}$$

while squaring the right-hand side of the equality gives

$$(\mathbf{p}'_{e})^2 = \mathbf{p}'_{e} \cdot \mathbf{p}'_{e} = -m_e^2 c^2. \tag{2.88}$$

Equating the two sides of the equation, we arrive at

$$2\mathbf{p}_{\gamma} \cdot \mathbf{p}_{e} - m_e^2 c^2 = -m_e^2 c^2 \quad \rightarrow \quad \mathbf{p}_{\gamma} \cdot \mathbf{p}_{e} = 0. \tag{2.89}$$

However, in the rest frame of the electron, where $\mathbf{p}_{e} = (m_e c, 0, 0, 0)^{\mathrm{T}}$ and $\mathbf{p}_{\gamma} = (E/c, \boldsymbol{p})^{\mathrm{T}}$, this means that the photon energy must be zero, *i.e.*, there is no photon.

Exercises

1. S and S' reference frames that are moving relative to one another with a velocity V in the x-direction measure the velocities of an object, also moving along the x-direction as v and v', respectively. Use the Lorentz transformations to derive the relationship between the velocities v and v'.
2. Flat Euclidean 2D space can be represented in terms of Cartesian coordinates (x, y), as well as in terms of polar coordinates (r, θ):

$$x = r\cos\theta, \qquad y = r\sin\theta. \tag{2.90}$$

   a. Write down the line element in Cartesian coordinates and the metric g.
   b. Write down the line element in polar coordinates and the metric g.
3. Consider 3D spherical coordinates.
   a. Write down the line element and the Euclidean metric g.
   b. Given the contravariant vector $x^i = (1, r, 0)^{\mathrm{T}}$, find x_i.
   c. Given the covariant vector $y_i = (1, -r^2, \cos^2\theta)$, find y^i.
4. Suppose that a particle of mass m_1 and three-velocity $\boldsymbol{v}_1$ collides with another particle at rest with mass m_2 and is absorbed by it.
   a. Use conservation of four-momentum to compute the three-velocity $\boldsymbol{v}$ of the composite particle.
   b. From the previous result, recover the Newtonian limit for $v_1 \ll c$.
   c. Use conservation of four-momentum to compute the mass m of the composite particle.
   (Hint: $m \neq m_1 + m_2$. Conservation of four-momentum demands that some of the "energy of motion" be converted into rest-mass energy.)
5. A photon of wavelength λ traveling to the left along the x-axis hits a stationary electron of mass m_e and scatters off it with a wavelength λ' at an angle θ in (x, z) plane (as measured in the frame of reference in which the electron was originally at rest). Use conservation of four-momentum to compute the scattering wavelength λ'.
   (Hint: Recall that a photon with frequency ν has a measured energy $E = h\nu$ and spatial momentum of magnitude $|\mathbf{p}| = h\nu/c$. Start with the equation of four-momentum conservation written in the form $\mathbf{p}_e + \mathbf{p}_\lambda = \mathbf{p}'_e + \mathbf{p}'_\lambda$ and recall that $\mathbf{p} \cdot \mathbf{p} = -m^2c^2$.)
6. Find another frame of reference in Example 1 above and compute what the observer observes for E.
7. Consider Fig. 2.4.
   a. What does the path of a particle following the red path look like in space?
   b. What does the path of a particle following the blue path look like in space? How fast is the particle moving and in which direction? What happens at the intermediate point C?

c. What does the path of a particle following the purple path look like in space? About how fast is the particle moving and in which direction?

d. Plot the path of a particle moving from the origin to the left at $v = c/3$ for $ct = 2$ m and then stops for $ct = 1$ m and then to the right at $v = 2c/3$ for $ct = 1$ m. Compute the proper time $d\tau$ measured by an experimenter traveling that path.

CHAPTER 3

The Equivalence Principle

No amount of experimentation can ever prove me right; a single experiment can prove me wrong.

Albert Einstein

The Big Picture: In this chapter, we describe the equivalence between the constant gravitational field and the constant acceleration. The Equivalence Principle ultimately leads us to view gravity as geometric effects, rather than a force.

3.1. Newtonian gravity is inconsistent with special relativity

Newtonian gravity is based on Newtonian mechanics, which includes that, in principle, a material object can move with arbitrarily large speeds. It assumes that the gravitational interaction between two masses is instantaneous, depending on the positions of the two masses at *simultaneous instants* of time: $\boldsymbol{F}_{12}(t)$, the force that m_2 exerts on m_1 at time t has a magnitude,

$$F_{12}(t) = \frac{Gm_1 m_2}{|\boldsymbol{r}_1(t) - \boldsymbol{r}_2(t)|^2}. \tag{3.1}$$

However, this relation cannot hold in all inertial frames because different observers in different frames will identify different events as simultaneous. Breakdown of simultaneity of events directly contradicts Newtonian notion of absolute time.

3.2. The equivalence of gravitational and inertial mass

When considering the other known basic interactions of nature – electromagnetism, the strong interaction, and the weak interaction – one can identify separately a charge q, which dictates how much the interaction affects some particle, and the *inertial mass* m_{I}, which dictates how difficult it is to affect. For example, in the context of electrostatics, a particle in a constant electric field $\boldsymbol{E}$ feels an electric force $\boldsymbol{F} = q\boldsymbol{E}$, so that $\boldsymbol{F} = m_{\mathrm{I}}\boldsymbol{a}$ reduces to

$$q\boldsymbol{E} = m_{\mathrm{I}}\boldsymbol{a} \quad \rightarrow \quad \boldsymbol{a} = \left(\frac{q}{m_{\mathrm{I}}}\right)\boldsymbol{E}. \tag{3.2}$$

Since the ratio q/m_{I} is different for different types of particles, these different particles will be accelerated differently in the same electric field.

Relativity and Cosmology
https://doi.org/10.1016/B978-0-44-323542-9.00011-0

In a similar fashion, a particle inserted into a constant gravitational field $\boldsymbol{g}$ can be said to feel a force $\boldsymbol{F} = m_G \boldsymbol{g}$, where m_G, the *gravitational mass*, plays the role of the charge, so that

$$m_G \boldsymbol{g} = m_I \boldsymbol{a} \quad \rightarrow \quad \boldsymbol{a} = \left(\frac{m_G}{m_I}\right)\boldsymbol{g}. \tag{3.3}$$

This looks very similar to the preceding equation, but there is one absolutely crucial difference: it is experimentally established to a very high precision that $m_G = m_I$, *i.e.*, the gravitational mass, which determines the magnitude of the gravitational force – how hard the particle is pulled – is *precisely* equal to the inertial mass, which determines how difficult the particle is to pull. This implies that every particle inserted into a gravitational field will experience the same acceleration.

Motivated by this observation, Einstein asked a very fundamental question: *Given that every particle experiences the same gravitational effects, how do we really know that there is a force called gravity?* Addressing this issue led to his version of the *Equivalence Principle*, which is the cornerstone of all Einstein's theory of gravity:

The Equivalence Principle: *The laws of physics have the same form in every uniformly accelerated laboratory as they do in an unaccelerated laboratory in a uniform gravitational field.*

The basic idea of the Equivalence Principle is very simple, albeit profound:

- Suppose that you are standing in a rocket in outer space in what is (essentially) an inertial frame, so that you feel no gravity, and that you are holding a cannonball in one hand. If you release the cannonball with zero velocity, it will not move away from you.
- Suppose now that your rocket is accelerating upward at a uniform rate $\boldsymbol{a}$. In this case, if you release the cannonball, it will move downward because of the rocket's upwards acceleration. Moreover, if the insides of the rocket constitute a vacuum, so that there is no air resistance, a feather released from the other hand will fall downward in exactly the same fashion as the cannonball.
- Exactly the same result would be obtained if, instead, the rocket were situated in a constant gravitational field with $\boldsymbol{g} = \boldsymbol{a}$. If the windows of the rocket were covered, you could not, in principle, determine whether you were perceiving the effects of a constant acceleration or the effects of a constant gravitational field.

3.2.1 Light must also feel gravity

The Equivalence Principle implies that light must also fall downward in a gravitational field! In empty space, a light ray will move in a straight line in an inertial frame. Suppose, however, that you are in a uniformly accelerated frame and looking at the light. If the light is seen to move horizontally in an inertial frame, but you are being accelerated upward in the vertical direction, in your frame of reference the light will appear to be accelerated downward. Since the effects of a uniform acceleration are identical to the effects of a uniform gravitational field, it follows that a constant gravitational field will also cause a light ray to accelerate downward.

3.2.2 Clocks in a uniform gravitational field

Consider a rocket of length h that is being accelerated upward at a constant rate $\boldsymbol{g}$. Suppose then that an observer at the head of the rocket beams a pulse of light toward the tail of the rocket every $\Delta\tau$. What is the time Δt that an observer at the tail of the rocket would measure between the successive pulses that she/he receives?

In addressing this problem, it is easiest – albeit not essential – to express everything from the viewpoint of an inertial frame. For simplicity, it is also convenient to consider the limit that $(v/c)^2$ and $(gh/c^2)^2$ are both very small, which implies that one can address this question using ordinary Newtonian mechanics, neglecting time dilation and other subtleties. Again, this is an assumption that can be relaxed in principle. In this approximation, assuming that the rocket is accelerated in the x-direction, the locations of the head and tail satisfy

$$x_H(t) = h + \frac{1}{2}gt^2 \quad \text{and} \quad x_T(t) = \frac{1}{2}gt^2. \tag{3.4}$$

Suppose, now, that two successive pulses leave the head of the rocket at times $t = 0$ and $t = \Delta\tau$ and that they reach the tail of the rocket at times $t = t_1$ and $t = t_1 + \Delta t$. What one wants to compute is a connection between $\Delta\tau$ and Δt. (See Fig. 3.1.)

The first pulse will travel a distance

$$x_H(0) - x_T(t_1) = -c0 - (-ct_1) = ct_1, \tag{3.5}$$

in the allotted time t_1 and, similarly, the second pulse will travel a distance

$$x_H(\Delta\tau) - x_T(t_1 + \Delta t) = -c\Delta\tau - (-c)(t_1 + \Delta t) = c(t_1 + \Delta t - \Delta\tau). \tag{3.6}$$

The second pulse will travel a *shorter* distance because of the acceleration.

If one inserts the expressions for $x_T(t)$ and $x_H(t)$ into the last two equations, we obtain:

$$x_H(0) - x_T(t_1) = h - \frac{1}{2}gt_1^2 = ct_1 \tag{3.7}$$

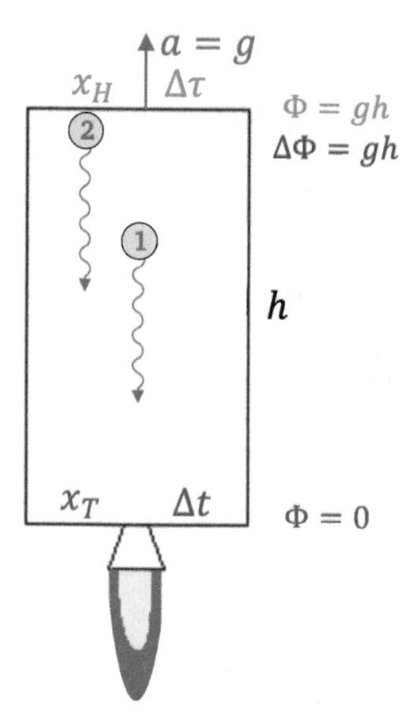

Figure 3.1 Pulses 1 and 2 traveling from the head (subscript *H*) to the tail (subscript *T*) of the rocket of length *h*, accelerating upward at the rate *g*. The time interval between when the two pulses were emitted ($\Delta\tau$) is longer than the time interval between when those pulses are detected (Δt). Therefore the clock at the head of the rocket runs faster.

and

$$\begin{aligned} x_H(\Delta\tau) - x_T(t_1 + \Delta t) &= h + \frac{1}{2}g(\Delta\tau)^2 - \frac{1}{2}g(t_1 + \Delta t)^2 \\ &= h + \frac{1}{2}g(\Delta\tau)^2 - \frac{1}{2}gt_1^2 - gt_1\Delta t - \frac{1}{2}(\Delta t)^2 \\ &= c(t_1 + \Delta t - \Delta\tau). \end{aligned} \tag{3.8}$$

If we assume that Δt is small, so that terms of order $(\Delta t)^2$ and $(\Delta\tau)^2$ may be ignored, it follows that

$$h - \frac{1}{2}gt_1^2 - gt_1\Delta t = c(t_1 + \Delta t - \Delta\tau). \tag{3.9}$$

What remains is to subtract Eq. (3.9) from Eq. (3.7) to obtain:

$$-gt_1\Delta t = c(\Delta t - \Delta\tau). \tag{3.10}$$

Given the assumptions that $(v/c)^2$ and $(gh/c^2)^2$ are being treated as negligible, one can approximate $h = ct_1 + (1/2)gt_1^2 \approx ct_1 \rightarrow t_1 = h/c$, after which it follows that

$$\frac{gh}{c^2}\Delta t = -\Delta t + \Delta\tau \quad \rightarrow \quad \Delta\tau = \Delta t\left(1 + \frac{gh}{c^2}\right) \quad \text{or} \quad \Delta t = \Delta\tau\left(1 - \frac{gh}{c^2}\right). \tag{3.11}$$

In other words, the interval Δt is shorter than $\Delta\tau$ by a factor $(1 - gh/c^2)$. What this implies is that, relative to a clock at the tail of the rocket, a clock at the head runs faster by a factor $(1 - gh/c^2)^{-1} \approx (1 + gh/c^2)$.

The crucial point – according to the Equivalence Principle – is that *if this result holds true for some object experiencing a constant acceleration, it must also hold for an object inserted into a constant gravitational field.* When one is treating gravity as the source of this effect, the quantity gh can be interpreted naturally as the change in gravitational potential between the head and tail of the rocket, *i.e.*, $\Delta\Phi = gh$, so that one infers that the rate at which a clock at the head of the rocket ticks is increased by a factor of $(1 + \Delta\Phi/c^2)$ relative to a clock at the tail of the rocket.

Note that $\Delta\tau$ corresponds to the proper time between ticks for a clock in a uniform gravitational field. The preceding equation can thus be reinterpreted as saying that *the interval Δt between the same two ticks measured by a clock at zero gravitational potential is*

$$\Delta t = \left(1 - \frac{\Phi}{c^2}\right)\Delta\tau. \tag{3.12}$$

This result has only been derived for the case of a constant gravitational field.

The basic assumption underlying general relativity, confirmed experimentally, is that this result also holds for nonuniform gravitational fields, so that one can write

$$\Delta t = \left(1 - \frac{\Phi(\mathbf{r})}{c^2}\right)\Delta\tau. \tag{3.13}$$

Alternatively, since it has been assumed throughout that Φ/c^2 is small compared to unity, one can equally well write:

$$\Delta\tau = \left(1 - \frac{\Phi(\mathbf{r})}{c^2}\right)^{-1}\Delta t \approx \left(1 + \frac{\Phi(\mathbf{r})}{c^2}\right)\Delta t. \tag{3.14}$$

3.3. A scientific fable

Once upon a time, there was a very smart scientist who, however, was convinced that the Earth is really flat. In particular, he believed that those portions of surface of the Earth that he had visited (basically the regions near the north Atlantic) were described correctly by a two-dimensional map where lines of latitude run horizontally and lines of longitude run vertically. He noted then the following sets of latitude and longitude for four places he had visited:

- London: 52° latitude, 0° longitude;

- Newfoundland: 52° latitude, 60° longitude;
- Ghana: 12° latitude, 0° longitude;
- Trinidad: 12° latitude, 60° longitude.

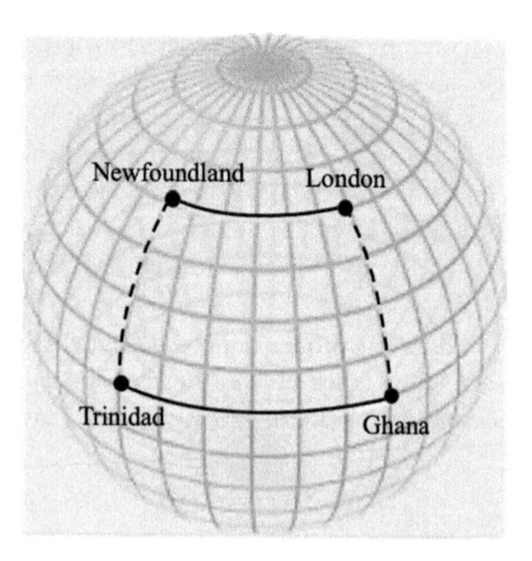

Figure 3.2 The distances along the surface of the Earth's sphere. Distances Newfoundland–Trinidad and London–Ghana (vertical lines) are along the lines of the same longitude (*great circles* of the same radius $r = R_E$, where R_E is the radius of the Earth. The distances Newfoundland–London and Trinidad–Ghana (horizontal lines), even though they cover the same angular distance (60° of longitude), are along the circles of different curvature, and are, therefore, different.

Obviously, he reasoned, the distances from London to Ghana and from Newfoundland to Trinidad must be equal; and, similarly, the distances from London to Newfoundland and from Ghana to Trinidad must again be the same. He was, in fact, correct in asserting that the London–Ghana and Newfoundland–Trinidad distances are equal, but he knew from experience that the distance from London to Newfoundland is substantially shorter than the distance between Trinidad and Ghana. (See Fig. 3.2.)

You and I would say this is because of the curvature of the Earth – it is really a near-sphere, rather than nearly flat – but our enterprising scientist simply would not accept the idea that the Earth is round. So what could he do? Eventually, he recognized that there *was*, in fact, a way for him to save his interpretation, namely by assuming that rulers (and other measuring apparati) measure lengths differently at different latitudes, at least in the East–West direction. They do not, however, look different to us, so we too and everything about us must change dimensions, at least in the East–West direction, as we move from higher to lower latitudes. How to account for this? He posited the experience of a special *field*, the value of which depends upon one's location and that is responsible essentially for dilating and shrinking objects at different latitudes and in different directions.

All this might sound crazy, but one could ask: Viewed simply as a tool to make experimental predictions, is his explanation any less accurate than the usual explanation, based on the idea that the Earth is round? If one allows for very large-scale measurements, the answer is 'yes': we know that if one travels far enough in any direction, one comes back to the point from which they began, whereas the flat-Earth interpretation would imply either that the Earth extends forever or that it has an edge. If, however, one restricts attention to comparatively local measurements, *there is no possible way in which one could prove experimentally that the flat-Earth model is incorrect!* Einstein would, no doubt, tell us that there is an important lesson hiding here.

3.4. Gravity as geometry

Stated in ordinary Newtonian language, the Equivalence Principle implies that the gravitational fields affect the rates at which clocks run and that all clocks are affected in exactly the same way. How much simpler, Einstein would argue, to say that the fashion in which clocks are measuring time is not changing, but that, because of the presence of mass, spacetime becomes curved, and that this curvature, like the curvature of the Earth, alters our expected notions of temporal distance. The presence of matter generates a gravitational potential, and this potential is manifested as spacetime curvature.

Two ticks on a stationary clock, for which $dx = dy = dz = 0$, are separated by a proper time $d\tau$, satisfying

$$d\tau = \left(1 + \frac{\Phi(\mathbf{r}, t)}{c^2}\right) dt, \tag{3.15}$$

which were derived assuming $v \ll c$ and $|\Phi(\mathbf{r}, t)| \ll c^2$. Since the gravitational field is weak,

$$d\tau^2 = \left(1 + \frac{\Phi(\mathbf{r}, t)}{c^2}\right)^2 dt^2 \approx \left(1 + \frac{2\Phi(\mathbf{r}, t)}{c^2}\right) dt^2. \tag{3.16}$$

dt is the proper time that would have been measured in the absence of a gravitational potential.

In the presence of a gravitational potential, however, the difference in the clock rates changes the spacetime interval to

$$ds^2 = -c^2 d\tau^2 = -\left(1 + \frac{2\Phi(\mathbf{r}, t)}{c^2}\right) c^2 dt^2 + dx^2 + dy^2 + dz^2. \tag{3.17}$$

3.5. Geometric derivation of Newtonian gravity

If this interpretation is correct, it must also yield the correct equation of motion for a particle moving in a gravitational potential, *i.e.*,

$$\frac{d^2\boldsymbol{r}}{dt^2} = -\nabla\Phi(\boldsymbol{r}, t). \tag{3.18}$$

In special relativity, the equation of motion for a free particle comes from the demand that the proper time for the particle's flat spacetime trajectory be extremal. Allowing for spacetime curvature, what one must presumably demand is that the proper time for the particle's trajectory in curved spacetime again be extremal. What this means is that one must extremize the action,

$$\tau_{AB} = \int_A^B d\tau = \int_A^B \left(-\frac{ds^2}{c^2}\right)^{1/2} = \int_A^B \left[\left(1 + \frac{2\Phi(\boldsymbol{r}, t)}{c^2}\right) dt^2 - \frac{1}{c^2}\left(dx^2 + dy^2 + dz^2\right)\right]^{1/2}. \tag{3.19}$$

If one chooses to parametrize the path by coordinate time t, so that it is characterized by specifying $x(t)$, $y(t)$, and $z(t)$, this becomes

$$\tau_{AB} = \int_{t(A)}^{t(B)} \left\{\left(1 + \frac{2\Phi(\boldsymbol{r}, t)}{c^2}\right) - \frac{1}{c^2}\left[\left(\frac{dx}{dt}\right)^2 + \left(\frac{dy}{dt}\right)^2 + \left(\frac{dz}{dt}\right)^2\right]\right\}^{1/2} dt. \tag{3.20}$$

The quantity in the square brackets is the square of the Newtonian velocity $\boldsymbol{v}^2$. Noting, however, that the entire analysis up until this point has been predicated on the assumption that $|\Phi| \ll c^2$ and $v \ll c$, one can expand the curly brackets in a series in powers of $1/c^2$, which yields, to lowest nontrivial order:

$$\tau_{AB} = \int_{t(A)}^{t(B)} \left[1 - \frac{2}{c^2}\left(\frac{1}{2}\boldsymbol{v}^2 - \Phi(\boldsymbol{r}, t)\right)\right]^{1/2} dt \approx \int_{t(A)}^{t(B)} 1 - \frac{1}{c^2}\left(\frac{1}{2}\boldsymbol{v}^2 - \Phi(\boldsymbol{r}, t)\right) dt. \tag{3.21}$$

Since one is interested in evaluating $\delta \int_A^B d\tau = 0$, one can ignore the constant term in the integrand and multiply the remaining terms by another constant, namely mc^2. This leads to the Variational Principle,

$$L\left(\frac{d\boldsymbol{r}}{dt}, \boldsymbol{r}\right) = \frac{m}{2}\left(\frac{d\boldsymbol{r}}{dt}\right)^2 - m\Phi(\boldsymbol{r}, t), \tag{3.22}$$

which yields the desired equation of motion, $d^2\boldsymbol{r}/dt^2 = -\nabla\Phi(\boldsymbol{r}, t)$.

The moral of the story is clear.

As a consequence of the Equivalence Principle, Newtonian gravity can be expressed completely in geometric terms. Newton would have said that "the presence of mass produces a gravitational potential Φ, which determines particle motion through the equation of motion $d^2\mathbf{r}/dt^2 = -\nabla\Phi(\mathbf{r}, t)$." The geometric viewpoint involves saying that "the presence of matter produces spacetime curvature described by

$$ds^2 = -c^2 d\tau^2 = -\left(1 + \frac{2\Phi(\mathbf{r}, t)}{c^2}\right) c^2 dt^2 + dx^2 + dy^2 + dz^2 \tag{3.23}$$

and particles move in this curved spacetime along paths of extremal proper time." *There is no such thing as a gravitational force. What we usually interpret as a gravitational force is simply spacetime curvature.*

CHAPTER 4

General relativity

Everything should be made as simple as possible, but not simpler.

Albert Einstein

The Big Picture: In this chapter, we are going to introduce the notation used in general relativity, define the metric, compare motion in flat and curved metrics, and derive the geodesic equation – an equivalent in curved spacetime to Newton's Second Law.

4.1. Notation and conventions

It is useful to codify several notations and conventions in general relativity:

- **4-vector**: $(ct, x, y, z) \rightarrow (x^0, x^1, x^2, x^3)$.
- **Indices convention**:
  - Roman letters (i, j, k, l, m, n) run from 1 to 3;
  - Greek letters $(\alpha, \beta, \gamma, \delta, \mu, \nu, \eta, \xi)$ run from 0 to 3.
- **Einstein summation** (summation over repeated indices): $\sum_{\beta=0}^{3} \frac{\partial x'^\alpha}{\partial x^\beta} v^\beta \equiv \frac{\partial x'^\alpha}{\partial x^\beta} v^\beta$.
- **Contravariant vector**: transforms as $A'^\alpha = \frac{\partial x'^\alpha}{\partial x^\beta} A^\beta$ (index is a superscript; column vector).
- **Covariant vector**: transforms as $A'_\alpha = \frac{\partial x^\beta}{\partial x'^\alpha} A_\beta$ (index is a subscript; row vector).
- **Tensors**:
  - **First rank** (one index):
    - Contravariant: $A'^\alpha = \frac{\partial x'^\alpha}{\partial x^\beta} A^\beta$.
    - Covariant: $A'_\alpha = \frac{\partial x^\beta}{\partial x'^\alpha} A_\beta$.
- **Tensors**:
  - **Second rank** (two indices):
    - Contravariant: $A'^{\alpha\beta} = \frac{\partial x'^\alpha}{\partial x^\xi} \frac{\partial x'^\beta}{\partial x^\nu} A^{\xi\nu}$.
    - Covariant: $A'_{\alpha\beta} = \frac{\partial x^\xi}{\partial x'^\alpha} \frac{\partial x^\nu}{\partial x'^\beta} A_{\xi\nu}$.
    - Mixed: $A'^\alpha_\beta = \frac{\partial x'^\alpha}{\partial x^\xi} \frac{\partial x^\nu}{\partial x'^\beta} A^\xi_\nu$.
  - **Nth rank** (N indices):
    - Mixed: $A'^{\alpha_1 \dots \alpha_s}_{\alpha_{s+1} \dots \alpha_N} = \frac{\partial x'^{\alpha_1}}{\partial x^{\beta_1}} \dots \frac{\partial x'^{\alpha_s}}{\partial x^{\beta_s}} \frac{\partial x^{\alpha_{s+1}}}{\partial x'^{\beta_{s+1}}} \dots \frac{\partial x^{\alpha_N}}{\partial x'^{\beta_N}} A^{\beta_1 \dots \beta_s}_{\beta_{s+1} \dots \beta_N}$.
  - **Operations with tensors:**
    - Addition: $A^{\alpha\beta}_{\xi\nu} + B^{\alpha\beta}_{\xi\nu} = C^{\alpha\beta}_{\xi\nu}$.

https://doi.org/10.1016/B978-0-44-323542-9.00012-2

- Subtraction: $A^{\alpha\beta}_{\xi\nu} - B^{\alpha\beta}_{kl} = D^{\alpha\beta}_{\xi\nu}$.
- Tensor product: $A^{\alpha\beta}_{\xi\nu} B^{\gamma\delta}_{\eta\psi} = G^{\alpha\beta\gamma\delta}_{\xi\nu\eta\psi}$.
- Contraction: $B^{\beta}_{\beta} = C$, or $A^{\alpha\beta}_{\beta\gamma} = H^{\alpha}_{\gamma}$ (summed over β).
- Inner product: $A^{\alpha\beta}_{\xi\nu} B^{\nu\gamma}_{\delta\eta} = P^{\alpha\beta\nu\gamma}_{\xi\nu\delta\eta} = K^{\alpha\beta\gamma}_{\xi\delta\eta}$.

• **Importance of tensors**: When written in tensor form, the equations of motion are invariant under appropriately defined transformation:
  - **Newtonian mechanics**: 3-vector (x^1, x^2, x^3) is invariant under Galilean transformation.
  - **Special relativity**: 4-vector (x^0, x^1, x^2, x^3) is invariant under Lorentz transformation.
  - **General relativity**: 4-vector (x^0, x^1, x^2, x^3) is invariant under general metric transformation.

- **Invariants**: scalars that are the same in all coordinate systems.
- When writing out equations describing the geometry of spacetime, the locations of the indices *must* be respected, *i.e.*, one must be careful to make sure that some are superscripts and others are subscripts. (Even in flat spacetime written in Cartesian coordinates, the quantities A^{α} and A_{α} are not in general equal numerically!)
- Repeated indices *always* occur once as a subscript and once as a superscript and are *always* summed: for this reason, they are sometimes called *dummy indices*. The act of summing over a repeated index is called *contraction*.
  • Quantities $g_{\alpha\beta}a^{\alpha}a^{\beta}$ and $g_{\mu\nu}a^{\mu}a^{\nu}$ mean exactly the same thing.
  • Quantities like $g_{\alpha\beta}A_{\alpha}$ are meaningless. If you see one, someone has made a mistake!
- Indices that are not summed are called *free indices*. They must balance on both sides of the equation.
  • The equation $g^{\alpha\beta}g_{\beta\gamma} = \delta^{\alpha}_{\gamma}$ is meaningful, but $g^{\alpha\beta}g_{\beta\xi} = \delta^{\alpha}_{\gamma}$ is not.
- For consistency of notation, $g_{\alpha\beta}g^{\beta\gamma} = \delta^{\gamma}_{\alpha} \equiv g^{\gamma}_{\alpha}$.

4.2. General coordinate transformation

How do vectors transform under a general coordinate transformation? For a general coordinate transformation from x^{α} to x'^{α}, one can write

$$x^{\alpha} = x^{\alpha}(x'^{\beta}) = x^{\alpha}(\mathbf{x}'), \tag{4.1}$$

where the four functions, *e.g.*, $x^0(\mathbf{x}') = x^0(x'^0, x'^1, x'^2, x'^3)$, are arbitrary. By exploiting the chain rule,

$$dx^{\alpha} = \frac{\partial x^{\alpha}}{\partial x'^{\beta}} dx'^{\beta}, \tag{4.2}$$

one sees then that

$$ds^2 = g_{\alpha\beta} dx^{\alpha} dx^{\beta} = \left(g_{\alpha\beta} \frac{\partial x^{\alpha}}{\partial x'^{\mu}} \frac{\partial x^{\beta}}{\partial x'^{\nu}} \right) dx'^{\mu} dx'^{\nu} \equiv g'_{\mu\nu}\, dx'^{\mu} dx'^{\nu}, \tag{4.3}$$

where $g_{\alpha\beta}$ is the metric tensor associated with coordinates x^{α}, and $g'_{\mu\nu}$

$$g'_{\mu\nu}(\mathbf{x}') \equiv g_{\alpha\beta}(\mathbf{x}) \frac{\partial x^{\alpha}}{\partial x'^{\mu}} \frac{\partial x^{\beta}}{\partial x'^{\nu}}, \tag{4.4}$$

is the metric tensor associated with coordinates x'^{α}. This is the general expression for ds^2 in any spacetime, flat or curved.

Given that the displacement $d\mathbf{x}$ is an example of a vector, it must be true in general that vectors transform like $d\mathbf{x}$ under coordinate transformations, *i.e.*, in accord with the chain rule.

Under a general coordinate transformation $x^{\alpha} = x^{\alpha}(x'^{\beta})$, a *contravariant vector* A^{α}, defined with its index as a superscript, transforms as

$$A^{\beta} = \frac{\partial x^{\beta}}{\partial x'^{\gamma}} A'^{\gamma}. \tag{4.5}$$

The quantity $A_{\alpha} = g_{\alpha\beta} A^{\beta}$ is defined to be a *covariant* vector and, given the rule for the transformation of the metric $g_{\alpha\beta}$ in Eq. (4.4), it follows that A_{α} transforms as

$$\begin{aligned} A_{\alpha} &= g_{\alpha\beta} A^{\beta} = g_{\alpha\beta} \frac{\partial x^{\beta}}{\partial x'^{\gamma}} A'^{\gamma} = \frac{\partial x'^{\mu}}{\partial x^{\alpha}} \frac{\partial x'^{\nu}}{\partial x^{\beta}} g'_{\mu\nu} \frac{\partial x^{\beta}}{\partial x'^{\gamma}} A'^{\gamma} = \frac{\partial x'^{\nu}}{\partial x^{\beta}} \frac{\partial x^{\beta}}{\partial x'^{\gamma}} \frac{\partial x'^{\mu}}{\partial x^{\alpha}} g'_{\mu\nu} A'^{\gamma} \\ &= \delta^{\nu}_{\gamma} \frac{\partial x'^{\mu}}{\partial x^{\alpha}} g'_{\mu\nu} A'^{\mu} = \frac{\partial x'^{\mu}}{\partial x^{\alpha}} g'_{\mu\gamma} A'^{\gamma} = \frac{\partial x'^{\mu}}{\partial x^{\alpha}} A'_{\mu}, \end{aligned} \tag{4.6}$$

after using Eq. (4.4) and the identity $\delta^{\nu}_{\gamma} = \frac{\partial x'^{\nu}}{\partial x^{\beta}} \frac{\partial x^{\beta}}{\partial x'^{\gamma}}$.

Under a general coordinate transformation $x'^{\alpha} = x'^{\alpha}(x^{\beta})$, a *covariant* vector A_{α}, defined with its index as a subscript, transforms as

$$A_{\alpha} = \frac{\partial x'^{\mu}}{\partial x^{\alpha}} A'_{\mu}. \tag{4.7}$$

If $A_\alpha = g_{\alpha\beta}A^\beta$, then $A^\alpha = g^{\alpha\beta}A_\beta$. Indeed, it follows from the definition of the inverse metric,

$$g^{\alpha\beta}g_{\beta\gamma} = \delta^\alpha_\gamma, \tag{4.8}$$

that

$$g^{\alpha\beta}A_\beta = g^{\alpha\beta}g_{\beta\gamma}A^\gamma = \delta^\alpha_\gamma A^\gamma = A^\alpha. \tag{4.9}$$

In a similar fashion, one can define an object like $A^{\alpha\beta}$ as a quantity that transforms as a contravariant vector on each of its two indices, $A_{\alpha\beta}$ as a quantity that transforms as a covariant vector on each of its two indices, and A^β_α as a quantity that transforms as a covariant vector with respect to its index α and contravariant with respect to its index β.

One can then use the metric $g_{\alpha\beta}$ to *raise* and the metric $g^{\alpha\beta}$ to *lower indices*, so that

$$A_{\alpha\beta} = g_{\alpha\nu}A^\nu_\beta, \qquad A^{\alpha\beta} = g^{\alpha\nu}A^\beta_\nu. \tag{4.10}$$

4.3. Spacetime coordinates and their metric tensors

It is important to understand that, when characterizing spacetime geometry, coordinates are not fundamental. They are simply a useful way of labeling different spacetime points. Changing the choice of coordinates used to describe the spacetime does not change the spacetime geometry. However, for a particular problem, a certain set of coordinates often seems particularly natural. For example, when studying central-force problems, it is generally easier to use spherical polar coordinates than Cartesian coordinates.

In flat spacetime, there exists a special class of observers with respect to whom the law of physics look especially simple, namely those in inertial frames. However, the fact that such inertial frames exist is a consequence of the special symmetries of flat spacetime. When considering a generic curved spacetime, without special symmetries, there will not exist a particular set of coordinate systems in which the general laws of physics look especially simple. What this means is that when considering physics in a curved spacetime, one really needs to formulate the general laws of physics in terms of arbitrary coordinates. A discussion of how this is done is the focus of this chapter.

Example 1: As a simple example of a coordinate transformation, one can consider how the standard expression for distance in flat spacetime,

$$ds^2 = -c^2 dt^2 + dx^2 + dy^2 + dz^2, \tag{4.11}$$

or, more generally written,

$$ds^2 = \eta_{\alpha\beta} dx^\alpha dx^\beta, \tag{4.12}$$

with $dx^\alpha = (cdt, dx, dy, dz)$ or $\mathbf{x} = x^\alpha = (ct, x, y, z)$, and

$$\eta_{\alpha\beta} = \begin{pmatrix} -1 & 0 & 0 & 0 \\ 0 & 1 & 0 & 0 \\ 0 & 0 & 1 & 0 \\ 0 & 0 & 0 & 1 \end{pmatrix} \tag{4.13}$$

is changed if one transforms to spherical polar coordinates $dx'^\alpha = (cdt, dr, d\theta, d\phi)$, or $\mathbf{x}' = x'^\alpha = (ct, r, \theta, \phi)$, setting

$$\begin{aligned} x^0(x'^\alpha) &= ct(ct, r, \theta, \phi) = ct, \\ x^1(x'^\alpha) &= x(ct, r, \theta, \phi) = r \sin\theta \cos\phi, \\ x^2(x'^\alpha) &= y(ct, r, \theta, \phi) = r \sin\theta \sin\phi, \\ x^3(x'^\alpha) &= z(ct, r, \theta, \phi) = r \cos\theta. \end{aligned} \tag{4.14}$$

After computing the requisite differentials,

$$dx^\alpha = \frac{\partial x^\alpha}{\partial x'^\beta} dx'^\beta, \tag{4.15}$$

to obtain

$$\begin{aligned} cdt &= cdt, \\ dx &= dr \sin\theta \cos\phi + r \cos\theta \cos\phi - r \sin\theta \sin\phi, \\ dy &= dr \sin\theta \sin\phi + r \cos\theta \sin\phi + r \sin\theta \cos\phi, \\ dz &= dr \cos\theta - r \sin\theta \, d\theta, \end{aligned} \tag{4.16}$$

one can show that, given this coordinate transformation, the expression for ds^2 becomes

$$ds^2 = -c^2 dt^2 + dr^2 + r^2 d\theta^2 + r^2 \sin^2\theta \, d\phi^2, \tag{4.17}$$

or, again, more generally,

$$ds^2 = \eta'_{\alpha\beta} dx'^{\alpha} dx'^{\beta}, \tag{4.18}$$

with $dx'^{\alpha} = (cdt, dr, d\theta, d\phi)$ and

$$\eta'_{\alpha\beta} = \begin{pmatrix} -1 & 0 & 0 & 0 \\ 0 & 1 & 0 & 0 \\ 0 & 0 & r^2 & 0 \\ 0 & 0 & 0 & r^2 \sin^2\theta \end{pmatrix}. \tag{4.19}$$

This looks very different from the expression for $\eta_{\alpha\beta}$. However, the physical notion of distance is unchanged and the geometry defined by this notion of distance is exactly the same. All that one has done has been to label spacetime coordinates in a different way.

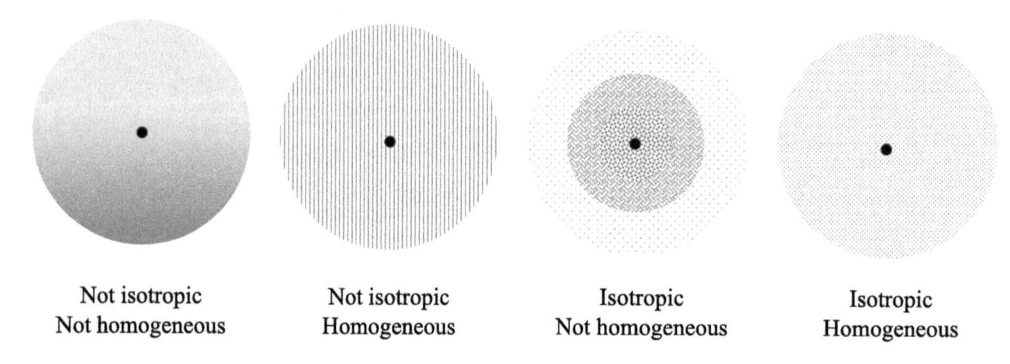

Figure 4.1 Homogeneity and isotropy for an observer located at the central dot.

4.3.1 Symmetries of a metric tensor

Since switching indices α and β does not change Eq. (4.3), the spacetime metric tensor is *always symmetric.* Furthermore, if one invokes the Cosmological Principle and assumes that the Universe is isotropic and homogeneous (Fig. 4.1) – as is the case in flat Euclidean space and the Minkowski and the Friedmann–Lemaître–Robertson–Walker spacetime metrics – the metric tensor in such a space will necessarily be *diagonal.*

Symmetry of the metric tensor (4×4 matrix) means that instead of 16, there are only 10 independent components. Since there are four arbitrary functions involved in these coordinate transformations in Eq. (4.3) – $x^{\alpha}(x'^0, x'^1, x'^2, x'^3)$ – there are only $10 - 4 = 6$ independent degrees of freedom associated with the spacetime metric $g_{\alpha\beta}$.

4.4. Metric tensors

4.4.1 Flat Euclidean space

Our common sense has taught us to think in terms of a flat *space* metric (Euclidean space), where parallel lines never cross and angles in a triangle always sum up to 180°, thus strongly reinforcing our Newtonian (incorrect!) notion of absolute space.

In this formulation, the invariant line element in Cartesian coordinates of a 3D *Euclidean space* (x^1, x^2, x^3) is

$$ds^2 = (dx^1)^2 + (dx^2)^2 + (dx^3)^2 \tag{4.20}$$

and space is assumed to be flat. Another way to write this is

$$ds^2 = \delta_{ij} dx^i dx^j, \tag{4.21}$$

where δ_{ij} is the Kronecker delta function ($\delta_{ij} = 1$ if $i = j$, $\delta_{ij} = 0$ otherwise). The Euclidean space metric tensor for Cartesian coordinates is given by

$$\delta_{ij} = \begin{pmatrix} 1 & 0 & 0 \\ 0 & 1 & 0 \\ 0 & 0 & 1 \end{pmatrix}. \tag{4.22}$$

An invariant line element in an arbitrary coordinate system in flat space can be written in terms of Cartesian coordinates (change of variables) as

$$ds^2 = \delta_{ij} dx^i dx^j = \delta_{ij} \frac{\partial x^i}{\partial x'^k} \frac{\partial x^j}{\partial x'^l} dx'^k dx'^l \equiv p_{kl} dx'^k dx'^l, \tag{4.23}$$

where p_{kl} is the space metric of the new coordinate system.

4.4.2 Flat Minkowski spacetime

Special relativity introduces 4-vectors in flat *Minkowski spacetime* $x^\alpha = (x^0, x^1, x^2, x^3) = (ct, x, y, z)$:

$$ds^2 = \eta_{\alpha\beta} dx^\alpha dx^\beta, \tag{4.24}$$

where $\eta_{\alpha\beta}$ is the flat *Minkowski spacetime metric tensor*:

$$\eta_{\alpha\beta} = \begin{pmatrix} -1 & 0 & 0 & 0 \\ 0 & 1 & 0 & 0 \\ 0 & 0 & 1 & 0 \\ 0 & 0 & 0 & 1 \end{pmatrix}. \tag{4.25}$$

It should be reiterated that the equation above specifies a Minkowski (flat) spacetime metric tensor *corresponding to coordinates* $x^\alpha = (ct, x, y, z)$. Any other choice of coordinates in which the flat spacetime can be expressed will require a coordinate transformation and another corresponding metric tensor.

For instance, if flat Minkowski spacetime is characterized by spherical polar coordinates $x'^\alpha = (ct, r, \theta, \phi)$, the metric tensor takes the form

$$\eta'_{\alpha\beta} = \begin{pmatrix} -1 & 0 & 0 & 0 \\ 0 & 1 & 0 & 0 \\ 0 & 0 & r^2 & 0 \\ 0 & 0 & 0 & r^2 \sin^2\theta \end{pmatrix}. \tag{4.26}$$

Expanding Flat Spacetime: The metric tensor, called the *Friedmann–Lemaître–Robertson–Walker metric tensor*, for a flat, homogeneous, and isotropic spacetime that is expanding in its spatial coordinates by a *scale factor* $a(t)$ is obtained from the Minkowski metric by scaling the spatial coordinates by $a^2(t)$:

$$g_{\alpha\beta} = \begin{pmatrix} -1 & 0 & 0 & 0 \\ 0 & a^2(t) & 0 & 0 \\ 0 & 0 & a^2(t) & 0 \\ 0 & 0 & 0 & a^2(t) \end{pmatrix}. \tag{4.27}$$

The equation above is for the expanding Minkowski metric in Cartesian coordinates.

4.4.3 Curved spacetime

When considering curved spacetimes, distance can be defined operationally in the same fashion as in flat spacetime, namely by: (1) setting up a system of coordinates x^α with which to label different spacetime points; (2) identifying dx^α as a separation between nearby pairs of points; and (3) specifying a *rule* that defines what is meant by the magnitude of the distance between two nearby points.

Specifying the rule amounts to defining the invariant line element in terms of a general (possibly curved) *covariant spacetime metric tensor* $g_{\alpha\beta}$.

The invariant line element in a curved spacetime described by a *covariant spacetime metric tensor* $g_{\alpha\beta}$ is

$$ds^2 = g_{\alpha\beta} dx^\alpha dx^\beta. \tag{4.28}$$

The *contravariant spacetime metric tensor* $g^{\alpha\beta}$ is simply an inverse of the *covariant spacetime metric tensor* $g_{\alpha\beta}$, so that

$$g^{\alpha\beta} g_{\beta\nu} = \delta^\alpha_\nu. \tag{4.29}$$

This implies that whenever the metric tensor is diagonal: $g^{00} = (g_{00})^{-1}$, $g^{11} = (g_{11})^{-1}$, $g^{22} = (g_{22})^{-1}$, $g^{33} = (g_{33})^{-1}$.

As with the Euclidean and Minkowski metric tensors, since indices α and β can be interchanged without changing the equation, both the covariant and the contravariant general spacetime metric tensors are *symmetric*.

4.5. Freely falling frames

In nonexpanding flat spacetime, one can always find a coordinate transformation so that the spacetime metric $g_{\alpha\beta}$ can be transformed to one with constant coefficients $\eta_{\alpha\beta}$ in Eq. (4.25). However, in curved spacetime, this is not true: $g_{\alpha\beta}$ has 10 independent components (due to symmetry), but one can only implement 4 coordinate transformations. Nevertheless, it is always possible to choose a set of coordinates so that, *at any given point* $\mathbf{x}_{(0)}$,

$$g_{\alpha\beta}(\mathbf{x}_{(0)}) = \eta_{\alpha\beta}, \qquad \text{and} \qquad \left.\frac{\partial g_{\alpha\beta}}{\partial x^\gamma}\right|_{\mathbf{x}=\mathbf{x}_{(0)}} = 0, \tag{4.30}$$

i.e., the metric looks like flat spacetime and its first derivative vanishes.

The fact is really a restatement, in part, of the Equivalence Principle. The Equivalence Principle says that, at least locally, a gravitational field is equivalent to a uniform acceleration and that, by moving with the correct acceleration, local gravitational effects can be transformed completely away. In so-called *freely falling frames of reference*, there are no observable gravitational effects, at least locally. Since general relativity says that gravity is really spacetime curvature, it must therefore be true that one can find a coordinate frame in which, at least locally, spacetime looks flat. As was shown earlier in Eq. (3.17), in an appropriate Newtonian limit, the first spatial derivative of the metric yields the gravitational force $\nabla\Phi$. One must, therefore, expect that, at any given spacetime point, the first derivative of the metric can be made to vanish. In general, it is *not* possible to make the second derivative of the metric vanish. This is a mathematical statement of the fact that one cannot eliminate the effects of a nonuniform gravitational field, which leads to tidal forces.

To actually show that there exist coordinates satisfying these conditions, it suffices to count the number of degrees of freedom involved in a general coordinate transformation at a point, say $\mathbf{x}'(0) = x'^{\alpha}_{(0)}$. A general coordinate transformation can be expanded in Taylor series as

$$
\begin{aligned}
x^{\alpha}(\mathbf{x}') = {} & x^{\alpha}(\mathbf{x}'(0)) + \left(\frac{\partial x^{\alpha}}{\partial x'^{\beta}}\right)_{\mathbf{x}'(0)} \left(x'^{\beta} - x'^{\beta}_{(0)}\right) \\
& + \frac{1}{2}\left(\frac{\partial^2 x^{\alpha}}{\partial x'^{\beta}\partial x'^{\mu}}\right)_{\mathbf{x}'(0)} \left(x'^{\beta} - x'^{\beta}_{(0)}\right)\left(x'^{\mu} - x'^{\mu}_{(0)}\right) \\
& + \frac{1}{6}\left(\frac{\partial^3 x^{\alpha}}{\partial x'^{\beta}\partial x'^{\mu}\partial x'^{\nu}}\right)_{\mathbf{x}'(0)} \left(x'^{\beta} - x'^{\beta}_{(0)}\right)\left(x'^{\mu} - x'^{\mu}_{(0)}\right)\left(x'^{\nu} - x'^{\nu}_{(0)}\right) + \ldots .
\end{aligned} \tag{4.31}
$$

As noted above, under a general coordinate transformation, the metric $g_{\alpha\beta}$ is transformed by factors involving $\partial x^{\beta}/\partial x'^{\alpha}$; and similarly, the derivative of the metric $\partial g_{\alpha\beta}/\partial x^{\delta}$ will be transformed by factors involving $\partial^2 x^{\gamma}/\partial x'^{\alpha}\partial x'^{\beta}$. At the point $\mathbf{x}'(0)$, there are 16 numbers $(\partial x^{\beta}/\partial x'^{\alpha})_{\mathbf{x}'(0)}$ that can be adjusted. Ten of these can be used to set $g'_{\alpha\beta} = \eta_{\alpha\beta}$ and the remaining six can then be used to implement an arbitrary four-dimensional rotation (*i.e.*, a combination of three spatial rotations and three Lorentz boosts). As $g_{\alpha\beta}$ is symmetric, the derivative $(\partial g'_{\alpha\beta}/\partial x'^{\gamma})_{\mathbf{x}'(0)}$ involves the specification of 40 numbers (10 $g'_{\alpha\beta}$ terms times 4 indices of x'^{γ}). However, there *are* 40 numbers $(\partial^2 x^{\gamma}/\partial x'^{\alpha}\partial x'^{\beta})_{\mathbf{x}'(0)}$ (4 indices in the numerator times 4 indices chosen for 2 locations in the denominator, with replacement: $4 \times C_2^{4+2-1} = 40$) that one can choose to fix. It follows that one can arrange for the first derivative of the metric to vanish at $\mathbf{x}'(0)$. However, the same does *not* work for the second derivative. There are 100 independent entities $(\partial^2 g'_{\alpha\beta}/\partial x'^{\mu}\partial x'^{\nu})_{\mathbf{x}'(0)}$ (10 coefficients $g'_{\alpha\beta}$ in the numerator times 4 indices chosen for 2 locations in the denominator, with replacement: $10 \times C_2^{4+2-1} = 100$) but only 80 quantities $(\partial^3 x^{\mu}/\partial x'^{\alpha}\partial x'^{\beta}\partial x'^{\gamma})_{\mathbf{x}'(0)}$, (4 indices in the numerator times 4 indices chosen for 3 locations in the denominator, with replacement: $4 \times C_3^{4+3-1} = 80$), so that there are, in general, 20 components of $(\partial^2 g'_{\alpha\beta}/\partial x'^{\mu}\partial x'^{\nu})_{\mathbf{x}'(0)}$ that cannot be transformed away. These represent the 20 independent degrees of freedom associated with spacetime curvature.

As spacetime can always be made to look flat in the neighborhood of any point, *i.e.*, $g'_{\alpha\beta} \to \eta_{\alpha\beta}$, for flat spacetime, one can again make a separation of directions in the neighborhood of the point into timelike, spacelike, and null directions. Just as for flat spacetime:

- $ds^2 > 0$ along *spacelike* directions;
- $ds^2 < 0$ along *timelike* directions;
- $ds^2 = 0$ along *null* directions.

Light rays move along null directions; material particles move along timelike directions. For particles, $-c^2 d\tau^2 = ds^2$ represents proper time. The proper time read by a clock traveling from the spacetime point A to a spacetime point B is

$$\tau_{AB} = \int_A^B d\tau = \int_A^B \sqrt{-\frac{ds^2}{c^2}} = \frac{1}{c}\int_A^B \sqrt{-g_{\alpha\beta}\,dx^\alpha dx^\beta}, \tag{4.32}$$

where the integral is along the worldline followed by the particle.

4.6. Area, volume, and four-volume

For a variety of problems, it is useful to know how to compute areas, volumes, and four-volumes. This will be illustrated here for the special case of a *diagonal metric*, for which

$$ds^2 = g_{00}(dx^0)^2 + g_{11}(dx^1)^2 + g_{22}(dx^2)^2 + g_{33}(dx^3)^2, \tag{4.33}$$

where the situation is comparatively simple. The important point is that, given this assumption, the four spacetime coordinates are orthogonal, so that the different directions that they represent can be treated independently. Consider, for example, an element of area on the (x^2, x^3) surface defined by the conditions $x^0 = const.$ and $x^1 = const.$ If the quantities dx^2 and dx^3 define *coordinate length* in the x^2 and x^3 directions, the *proper lengths* will be $d\ell^2 = \sqrt{g_{22}}dx^2$ and $d\ell^3 = \sqrt{g_{33}}dx^3$, and the element of two-dimensional area becomes

$$dA = d\ell^2 d\ell^3 = \sqrt{g_{22}g_{33}}\,dx^2 dx^3. \tag{4.34}$$

In a similar fashion, the element of three-dimensional volume on a $x^0 = const.$ surface is

$$dV = \sqrt{g_{11}g_{22}g_{33}}\,dx^1 dx^2 dx^3, \tag{4.35}$$

and the four-dimensional volume element becomes

$$d\Omega = \sqrt{-g_{00}g_{11}g_{22}g_{33}}\,dx^0 dx^1 dx^2 dx^3. \tag{4.36}$$

Here, the minus sign ensures that $d\Omega$ is real when applied to flat Minkowski spacetime (recall that the determinant of the metric $\eta_{\alpha\beta}$ equals -1).

Example 2: As a simple example, consider flat space in spherical polar coordinates, for which

$$ds^2 = -c^2 dt^2 + dr^2 + r^2 d\theta^2 + r^2 \sin^2\theta \, d\phi^2. \tag{4.37}$$

The diagonal metric tensor for the flat spacetime in spherical polar coordinates is given in Eq. (4.26). The coordinates are $x^\alpha = (ct, r, \theta, \phi)$. The area is then found to be

$$dA = d\ell^2 d\ell^3 = \sqrt{g_{22} g_{33}} dx^2 dx^3 = \sqrt{r^2 r^2 \sin^2\theta} \; d\theta^2 \; d\phi^3 = r^2 \sin\theta \; d\theta \; d\phi, \tag{4.38}$$

as the notion of surface area on an $r = const.$ sphere at fixed t. The volume is

$$dV = \sqrt{g_{11} g_{22} g_{33}} dx^1 dx^2 dx^3 = \sqrt{r^2 r^2 \sin^2\theta} \; dr \; d\theta \; d\phi = r^2 \sin\theta \; dr \; d\theta \; d\phi, \tag{4.39}$$

as the natural volume element for fixed time t.

4.7. Vectors in curved spacetime

In ordinary flat spacetime, vectors are defined as directed line segments connecting different points. However, this approach will not work in curved spacetimes, since the geodesic lines of shortest distance between two points depend on the ambient geometry. Physically, this poses no problem, since the 'useful' vectorial quantities, like momentum, velocity, and current, are intrinsically local in character, *i.e.*, they are objects that can be measured in a tiny spacetime region in the neighborhood of a single spacetime point.

Operationally, the natural way in which to define vectors is to exploit the fact that, in the infinitesimally small neighborhood around any spacetime point, spacetime looks 'nearly flat'. Physically, this is equivalent to saying that a tiny patch on the surface of a sphere looks almost like a patch of flat two-dimensional space. Mathematically, this follows from the fact that one can always choose coordinates for which, at any given point, $g_{\alpha\beta} = \eta_{\alpha\beta}$ and $\partial g_{\alpha\beta}/\partial x^\gamma$ vanishes (see Eq. (4.30) and the discussion that follows it). The key point is that, since the spacetime looks flat near the point, one can define infinitesimal vectors near that point in the same fashion as one does ordinary vectors in flat spacetime; and, given these infinitesimal vectors, one can construct other, more general vectors by multiplication with numbers and addition/subtraction using the ordinary flat-space rules. The mathematician would call this "defining vectors in the *tangent space*." A *vector field* is simply a rule for generating in a smooth fashion a vector at each spacetime point. It may be written as $\mathbf{A} = \mathbf{A}(\mathbf{x})$.

If vectors are to be useful, one needs to know what is meant by quantities like the scalar product $\mathbf{A} \cdot \mathbf{B}$. Operationally, they can be defined in the same fashion as in flat spacetimes.

In the neighborhood of any spacetime point, one can construct a basis of four vectors $\mathbf{e}_\alpha$ in terms of which any other vector A^α can be expressed as a linear combination, *i.e.*,

$$\mathbf{A}(\mathbf{x}) = A^\alpha(\mathbf{x})\mathbf{e}_\alpha(\mathbf{x}). \tag{4.40}$$

The quantities A^α correspond to the components of the vector in the basis $\mathbf{e}_\alpha$.

The scalar product of two vectors satisfies

$$\mathbf{A} \cdot \mathbf{B} = (A^\alpha \mathbf{e}_\alpha) \cdot (B^\beta \mathbf{e}_\beta) = (\mathbf{e}_\alpha \cdot \mathbf{e}_\beta) A^\alpha B^\beta, \tag{4.41}$$

and this quantity can be readily evaluated once one knows the scalar product between the different basis vectors.

In general, one could pick basis vectors in a more or less arbitrary fashion, but there are two 'natural' possibilities that are of special importance.

4.7.1 Coordinate bases

A coordinate basis is one in which the contravariant components of the displacement vector $d\mathbf{x}$ between two nearby points are just the coordinate separations dx^α, *i.e.*,

$$d\mathbf{x} = \frac{\partial \mathbf{x}}{\partial x^\alpha} dx^\alpha \equiv dx^\alpha \mathbf{e}_\alpha, \tag{4.42}$$

so that $\mathbf{e}_\alpha \equiv \partial \mathbf{x}/\partial x^\alpha$. With this definition,

$$ds^2 = d\mathbf{x} \cdot d\mathbf{x} = (dx^\alpha \mathbf{e}_\alpha)(dx^\beta \mathbf{e}_\beta) = (\mathbf{e}_\alpha \cdot \mathbf{e}_\beta) dx^\alpha dx^\beta = g_{\alpha\beta} dx^\alpha dx^\beta. \tag{4.43}$$

In a *coordinate basis*,

$$\mathbf{e}_\alpha \cdot \mathbf{e}_\beta = g_{\alpha\beta} \tag{4.44}$$

and

$$\mathbf{A} \cdot \mathbf{B} = g_{\alpha\beta} A^\alpha B^\beta. \tag{4.45}$$

Coordinate bases are often the most useful for performing computations.

4.7.2 Orthonormal bases

In an *orthonormal basis*, the basis vectors are defined to be of unit length and orthogonal to one another, so that

$$\mathbf{e}'_\alpha \cdot \mathbf{e}'_\beta = \eta_{\alpha\beta}, \tag{4.46}$$

where $\eta_{\alpha\beta}$ is the flat spacetime metric tensor. With this definition,

$$ds^2 = d\mathbf{x} \cdot d\mathbf{x} = (dx'^\alpha \mathbf{e}'_\alpha)(dx'^\beta \mathbf{e}'_\beta) = (\mathbf{e}'_\alpha \cdot \mathbf{e}'_\beta) dx'^\alpha dx'^\beta = \eta_{\alpha\beta} dx'^\alpha dx'^\beta. \tag{4.47}$$

Such a basis, therefore, corresponds physically to the type of basis that an observer carries along as he or she moves through spacetime. It is a local free-falling (local Lorentz) frame for which the underlying metric is the Minkowski metric $\eta_{\alpha\beta}$. The timelike vector $\mathbf{e}_0$ is the observer's four-velocity, and the three spatial vectors $\mathbf{e}_i$ are three unit vectors that define locally the axes of the observer's laboratory.

This type of basis is important because it is the components as defined in the basis that are actually measured in a physical experiment. In particular, if $\mathbf{e}_\alpha$ is an orthonormal basis appropriate for a particular observer and $\mathbf{p}$ the momentum of some particle being observed, for which

$$\mathbf{p} = p^\alpha \mathbf{e}'_\alpha, \tag{4.48}$$

then $E = cp^0$ is the observed energy and p^i are the components of the three-momentum.

Example 3: Suppose that one is considering flat spacetime in spherical polar coordinates, so that

$$ds^2 = -c^2 dt^2 + dr^2 + r^2 d\theta^2 + r^2 \sin^2\theta \, d\phi^2. \tag{4.49}$$

Both the coordinate and orthonormal bases will correspond to a collection of four vectors, each one directed in the t, r, θ, and ϕ directions. These correspond to the spacetime coordinates in spherical polar coordinates (ct, r, θ, ϕ). The non-zero elements of the diagonal spacetime metric are $g_{00} = -1$, $g_{11} = 1$, $g_{22} = r^2$, $g_{33} = r^2 \sin^2\theta$. The coordinate basis vectors satisfy

$$\mathbf{e}^0 = c\mathbf{dt}, \qquad \mathbf{e}^1 = \mathbf{dr}, \qquad \mathbf{e}^2 = \mathbf{d\theta}, \qquad \mathbf{e}^3 = \mathbf{d\phi}. \tag{4.50}$$

By contrast, the orthonormal basis vectors satisfy

$$\mathbf{e}'^0 = c\mathbf{dt}, \qquad \mathbf{e}'^1 = \mathbf{dr}, \qquad \mathbf{e}'^2 = r\mathbf{d\theta}, \qquad \mathbf{e}'^3 = r\sin\theta \, \mathbf{d\phi}. \tag{4.51}$$

The orthonormal vectors are 'more physical' in the sense that they have all the same dimensions of length, but it is often more convenient to work with the coordinate basis vectors.

Another example of this sort of behavior arises in the context of ordinary Lagrangian mechanics. If one chooses to write the kinetic energy $(1/2)m\boldsymbol{v}^2$ in spherical polar coordinates, one finds that the θ component of the canonical momentum $P_\theta = \partial L/\partial(d\theta/dt)$ satisfies $P_\theta = mr^2(d\theta/dt)$, whereas the physical momentum is, of course, $p_\theta = mr(d\theta/dt)$.

Example 4: Suppose that one is considering a curved spacetime for which

$$ds^2 = a^2(\eta)(-c^2 d\eta^2 + dr^2 + r^2 d\theta^2 + r^2 \sin^2\theta\, d\phi^2). \tag{4.52}$$

(As will be seen later, this corresponds to the line element appropriate for a spatially flat Friedmann cosmology, with a the so-called *scale factor* and temporal distance described in terms of *conformal time* η.) What are the coordinate and orthonormal basis vectors?
In this case, the only non-zero spacetime metric elements are $g_{00} = -a^2(\eta)$, $g_{11} = a^2(\eta)$, $g_{22} = a^2(\eta)r^2$, $g_{33} = a^2(\eta)r^2\sin^2\theta$. The coordinate basis vectors satisfy

$$\mathbf{e}^0 = c d\boldsymbol{\eta}, \qquad \mathbf{e}^1 = d\mathbf{r}, \qquad \mathbf{e}^2 = d\boldsymbol{\theta}, \qquad \mathbf{e}^3 = d\boldsymbol{\phi}, \tag{4.53}$$

whereas the orthonormal basis vectors are

$$\mathbf{e}'^0 = acd\boldsymbol{\eta}, \qquad \mathbf{e}'^1 = ad\mathbf{r}, \qquad \mathbf{e}'^2 = ard\boldsymbol{\theta}, \qquad \mathbf{e}'^3 = ar\sin\theta\, d\boldsymbol{\phi}. \tag{4.54}$$

An observer who is at rest in the sense that there is no motion in the r, θ, and ϕ directions must have a four-velocity that points completely in the η direction, so that the only non-vanishing component of the four-velocity u^α is u^0. The fact that $\mathbf{u} \cdot \mathbf{u} = -c^2$, derived in Eq. (2.60), therefore implies that

$$\mathbf{u} \cdot \mathbf{u} = g_{00}u^0u^0 = -a^2(\eta)u^0u^0 = -c^2, \tag{4.55}$$

so that $u^0 = c/a$ and $u_0 = g_{00}u^0 = ac$. It follows that the four-velocity of an observer at rest satisfies $\mathbf{u} = acd\boldsymbol{\eta}$, *i.e.*, that the four-velocity coincides with the orthonormal basis vector $\mathbf{e}'^0$.

Example 5: Suppose now that a particle of mass m is moving in the x-direction, so that it has a velocity $dx/d\eta = v$. What is its four-momentum p^α?

Since there is no motion in the y- or z-directions,

$$p^\alpha = m_0\left(\frac{cd\eta}{d\tau},\frac{dx}{d\tau},0,0\right)^{\mathrm{T}} = m_0\left(\frac{cd\eta}{d\tau},\frac{dx}{d\eta}\frac{d\eta}{d\tau},0,0\right)^{\mathrm{T}} = m_0\left(\frac{cd\eta}{d\tau},v\frac{d\eta}{d\tau},0,0\right)^{\mathrm{T}}. \tag{4.56}$$

The condition that $p_\alpha p^\alpha = -m_0^2c^2$, derived in Eq. (2.68), thus yields

$$-m_0^2c^2 = p_\alpha p^\alpha = g_{\alpha\beta}p^\beta p^\alpha = m_0^2c^2a^2\left(\frac{d\eta}{d\tau}\right)^2\left(-1+\frac{v^2}{c^2}\right) = -m_0^2c^2\left(\frac{d\eta}{d\tau}\right)^2\frac{a^2}{\gamma^2}, \tag{4.57}$$

where we used the spacetime metric elements $g_{00} = -a^2$ and $g_{11} = 1$, as well as the definition of the relativistic $\gamma = 1/\sqrt{1-v^2/c^2}$. Solving for $d\eta/d\tau$, we obtain

$$\frac{d\eta}{d\tau} = \frac{\gamma}{a}, \tag{4.58}$$

so that the elements of the four-momentum in Eq. (4.56) become

$$p^\alpha = m_0\left(\frac{c\gamma}{a},\frac{v\gamma}{a},0,0\right)^{\mathrm{T}} = \frac{\gamma m_0 c}{a}(1,\beta,0,0)^{\mathrm{T}}, \tag{4.59}$$

where the relativistic $\beta = v/c$. The energy measured by an observer 'at rest' (here, $\mathbf{u} = (c/a,0,0,0)^{\mathrm{T}}$) satisfies

$$E = -\mathbf{p}\cdot\mathbf{u} = -g_{\alpha\beta}p^\alpha u^\beta = \gamma m_0 ca(1,\beta,0,0)\cdot(c/a,0,0,0)^{\mathrm{T}} = \gamma m_0 c^2, \tag{4.60}$$

where, again, we used $g_{00} = -a^2$ and $g_{\alpha\beta}p^\alpha = p_\beta = -\gamma m_0 ca(1,\beta,0,0)$. This analysis could be performed in terms of so-called *cosmic time*, for which

$$ds^2 = -c^2dt^2 + a^2(t)(dr^2 + r^2d\theta^2 + r^2\sin^2\theta\, d\phi^2). \tag{4.61}$$

With these coordinates, the orthonormal basis vectors become

$$\mathbf{e}'^0 = cd\mathbf{t}, \qquad \mathbf{e}'^1 = ad\mathbf{r}, \qquad \mathbf{e}'^2 = ard\boldsymbol{\theta}, \qquad \mathbf{e}'^3 = ar\sin\theta\, d\boldsymbol{\phi}, \tag{4.62}$$

and the four-momentum of a particle moving $dx/d\eta = v$ or $dx/dt = (dx/d\eta)(d\eta/dt) = v/a$

$$p^\alpha = m_0\left(\frac{c}{\sqrt{1-v^2/c^2}},\frac{v}{\sqrt{1-v^2/c^2}},0,0\right)^{\mathrm{T}} = \gamma m_0 c(1,\beta,0,0)^{\mathrm{T}}. \tag{4.63}$$

After using $g_{00} = -1$, $g_{\alpha\beta}p^\alpha = p_\beta = -\gamma m_0 c(1,\beta,0,0)$, and the fact that the particle at rest will have $\mathbf{u} = (c,0,0,0)^{\mathrm{T}}$, the measured energy is again

$$E = -\mathbf{p}\cdot\mathbf{u} = -g_{\alpha\beta}p^\alpha u^\beta = \gamma m_0 c(1,\beta,0,0)\cdot(c,0,0,0)^{\mathrm{T}} = \gamma m_0 c^2. \tag{4.64}$$

4.8. Covariant derivative

Consider a vector **A** given in terms of its components along the basis vectors:

$$\mathbf{A} = A^{\alpha}\mathbf{e}_{\alpha}. \tag{4.65}$$

Differentiating the vector **A** using the Leibniz rule $(fg)' = f'g + g'f$, we obtain

$$\frac{\partial \mathbf{A}}{\partial x^{\alpha}} = \frac{\partial}{\partial x^{\alpha}}\left(A^{\beta}\mathbf{e}_{\beta}\right) = \frac{\partial A^{\beta}}{\partial x^{\alpha}}\mathbf{e}_{\beta} + A^{\beta}\frac{\partial \mathbf{e}_{\beta}}{\partial x^{\alpha}}. \tag{4.66}$$

In flat Cartesian coordinates, the basis vectors are constant, so the last term in the equation above vanishes. However, this is not the case in general curved spaces.

In general, the derivative in the last term will not vanish, and it will itself be given in terms of the original basis vectors:

$$\frac{\partial \mathbf{e}_{\beta}}{\partial x^{\alpha}} = \Gamma^{\nu}_{\alpha\beta}\mathbf{e}_{\nu}. \tag{4.67}$$

$\Gamma^{\nu}_{\alpha\beta}$ are called the *Christoffel symbols* [7] (or *affine connection*). They are defined in terms of a spacetime metric $g_{\alpha\beta}$:

$$\Gamma^{\nu}_{\alpha\beta} \equiv \frac{1}{2}g^{\nu\gamma}\left(g_{\alpha\gamma,\beta} + g_{\gamma\beta,\alpha} - g_{\alpha\beta,\gamma}\right). \tag{4.68}$$

Taking the curvature of the ambient manifold into account when taking derivatives of vectors or tensors yields a *covariant derivative*:

$$\begin{aligned} A_{\alpha;\beta} &\equiv A_{\alpha,\beta} - \Gamma^{\nu}_{\alpha\beta}A_{\nu}, \\ A^{\alpha}_{;\beta} &\equiv A^{\alpha}_{,\beta} + \Gamma^{\alpha}_{\beta\nu}A^{\nu}, \end{aligned} \tag{4.69}$$

where $A_{\alpha,\beta} \equiv \partial A_{\alpha}/\partial x^{\beta}$ and $A^{\alpha}_{,\beta} \equiv \partial A^{\alpha}/\partial x^{\beta}$. The covariant derivative is denoted with ; followed by an index, placed as a subscript of a vector for which the covariant derivative is computed.

For vectors A^{α} and A_{α} defined along a curve $x^{\beta} = x^{\beta}(\lambda)$, the *covariant derivatives along the curve* λ are found by multiplying Eq. (4.69) by dx^{β}/ds and summing over the index β to obtain

$$\frac{DA_\alpha}{D\lambda} \equiv \frac{dA_\alpha}{d\lambda} - \Gamma^\nu_{\alpha\beta}\frac{dx^\beta}{d\lambda}A_\nu, \qquad \frac{DA^\alpha}{D\lambda} \equiv \frac{dA^\alpha}{d\lambda} + \Gamma^\alpha_{\beta\nu}\frac{dx^\beta}{d\lambda}A^\nu. \tag{4.70}$$

A covariant derivative is a curved spacetime analog of the ordinary derivative in Cartesian coordinates in flat spacetime.

Principle of General Covariance: *All tensor equations valid in special relativity will also be valid in general relativity if:*

- *The Minkowski metric $\eta_{\alpha\beta}$ is replaced by a general curved metric $g_{\alpha\beta}$*
- *All partial derivatives are replaced by covariant derivatives: $, \to ;$*

Example 6:

$$\begin{aligned} d\tau^2 = -\eta_{\alpha\beta}dx^\alpha dx^\beta \quad &\to \quad d\tau^2 = -g_{\alpha\beta}dx^\alpha dx^\beta, \\ \eta_{\alpha\beta}u^\alpha u^\beta = -1 \quad &\to \quad g_{\alpha\beta}u^\alpha u^\beta = -1, \\ T^{\alpha\beta}_{\ \ ,\beta} = 0 \quad &\to \quad T^{\alpha\beta}_{\ \ ;\beta} = 0. \end{aligned} \tag{4.71}$$

4.9. Geodesic equation

In classical mechanics, Newton's Second Law states that a force imparts acceleration on the body it acts on:

$$m\frac{d^2\boldsymbol{x}}{dt^2} = \boldsymbol{F} = -\boldsymbol{\nabla}\Phi \qquad \to \qquad \frac{d^2\boldsymbol{x}}{dt^2} = -\frac{1}{m}\boldsymbol{\nabla}\Phi. \tag{4.72}$$

In the absence of forces, Newton's Second Law reduces to Newton's First Law:

$$\frac{d^2\boldsymbol{x}}{dt^2} = \boldsymbol{0}, \tag{4.73}$$

where $\boldsymbol{0} = (0, 0, 0)$. In flat 3D Euclidean space and flat Minkowski spacetime, this leads to straight lines $\boldsymbol{x} = \boldsymbol{s}t + \boldsymbol{b}$, where $\boldsymbol{s}$ is the slope vector and $\boldsymbol{b}$ is the constant vector (3-vector in 3D Euclidean space and 4-vector in Minkowski spacetime).

It is a fundamental assumption of general relativity that, in curved spacetimes, free particles (*i.e.*, particles feeling no non-gravitational effects) follow paths that extremize their proper interval ds. Such paths are called *geodesics*. Therefore generalizing Newton's laws of motion of a particle in the absence of forces, Eq. (4.73), to a general curved spacetime metric leads to the *geodesic equation*. The geodesic equation is the basic equation describing the motion of particles in curved spacetime.

We derive the geodesic equation here using the variational principle, leading to Lagrange's equations. Suppose the points x^i lie on a curve parametrized by λ, *i.e.*,

$$x^\alpha \equiv x^\alpha(\lambda), \qquad dx^\alpha = \frac{dx^\alpha}{d\lambda} d\lambda \tag{4.74}$$

and the distance between two points A and B is given by

$$s_{AB} = \int_A^B ds = \int_A^B \frac{ds}{d\lambda} d\lambda = \int_A^B \sqrt{g_{\alpha\beta} \frac{dx^\alpha}{d\lambda} \frac{dx^\beta}{d\lambda}} d\lambda \equiv \int_A^B L\left(\lambda, x, \frac{dx}{d\lambda}\right) d\lambda, \tag{4.75}$$

where L is the Lagrangian. The shortest path between the points A and B, the geodesic, is found by extremizing the path s_{AB}. This is done by standard tools of variational calculus that lead to Lagrange equations. We derive them here as a reminder.

Consider

$$G \equiv \int_A^B L\left(\lambda, x, \frac{dx}{d\lambda}\right) d\lambda. \tag{4.76}$$

Let $x = X(\lambda)$ be the curve extremizing G. Then, a nearby curve passing through A and B can be parametrized as $x = X(\lambda) + \varepsilon\eta(\lambda)$, such that $\eta(A) = \eta(B) = 0$. Extremizing Eq. (4.76) leads to

$$\begin{aligned}
\left.\frac{dG}{d\varepsilon}\right|_{\varepsilon=0} &= \int_A^B \left(\frac{\partial L}{\partial x}\eta + \frac{\partial L}{\partial \dot{x}}\dot{\eta}\right) d\lambda, && \text{where } \dot{x} \equiv \frac{dx}{d\lambda}, \quad \dot{\eta} \equiv \frac{d\eta}{d\lambda} \\
&= \int_A^B \frac{\partial L}{\partial x}\eta d\lambda + \int_A^B \frac{\partial L}{\partial \dot{x}}\dot{\eta} d\lambda. && \text{Now, integrate by parts} \\
&= \int_A^B \frac{\partial L}{\partial x}\eta d\lambda + \frac{\partial L}{\partial \dot{x}}\eta|_A^B - \int_A^B \frac{d}{d\lambda}\frac{\partial L}{\partial \dot{x}}\eta d\lambda \\
&= \int_A^B \eta\left[\frac{\partial L}{\partial x} - \frac{d}{d\lambda}\frac{\partial L}{\partial \dot{x}}\right] d\lambda = 0. && \text{Recall}: \eta(A) = \eta(B) = 0
\end{aligned} \tag{4.77}$$

However, the function η is arbitrary, so in order to have $dG/d\varepsilon|_{\varepsilon=0}$, the bracket in the integrand must vanish, and so we arrive at Lagrange's equations:

$$\frac{\partial L}{\partial x} - \frac{d}{d\lambda}\frac{\partial L}{\partial \dot{x}} = 0, \tag{4.78}$$

which can be extended to any number of phase-space coordinates:

$$\frac{\partial L}{\partial x^\alpha} - \frac{d}{d\lambda}\frac{\partial L}{\partial \dot{x}^\alpha} = 0. \tag{4.79}$$

After this little side-derivation, let us march on toward the geodesic equation.

We now introduce an *alternative Lagrangian:*

$$L = \frac{1}{2} g_{\gamma\delta} \dot{x}^{\gamma} \dot{x}^{\delta}. \tag{4.80}$$

One can, of course, use a more traditional form for the Lagrangian: $L = \sqrt{g_{\gamma\delta}\dot{x}^{\gamma}\dot{x}^{\delta}}$, but the mathematics is a lot cleaner with this choice (see Appendices A and B). This freedom comes from the fact that maximizing a square of a function is equivalent to maximizing the function itself (also, additive and multiplicative constants can be neglected).

After substituting Eq. (4.80) into Eq. (4.79), we have

$$\frac{1}{2} g_{\gamma\delta,\alpha} \dot{x}^{\gamma} \dot{x}^{\delta} - \frac{d}{d\lambda}\left(g_{\gamma\alpha}\dot{x}^{\gamma}\right) = 0, \tag{4.81}$$

where $g_{\gamma\delta,\alpha} \equiv \partial g_{\gamma\delta}/\partial x^{\alpha}$. After recognizing that

$$\frac{d}{d\lambda} g_{\gamma\alpha} = \frac{\partial g_{\gamma\alpha}}{\partial x^{\delta}} \dot{x}^{\delta} = g_{\gamma\alpha,\delta}\dot{x}^{\delta}, \tag{4.82}$$

we obtain

$$\frac{1}{2} g_{\gamma\delta,\alpha} \dot{x}^{\gamma} \dot{x}^{\delta} - g_{\gamma\alpha,\delta}\dot{x}^{\delta}\dot{x}^{\gamma} - g_{\gamma\alpha}\ddot{x}^{\gamma} =$$

$$\left(\frac{1}{2} g_{\gamma\delta,\alpha} - g_{\gamma\alpha,\delta}\right)\dot{x}^{\gamma}\dot{x}^{\delta} - g_{\gamma\alpha}\ddot{x}^{\gamma} = 0.$$

Multiplying by $g^{\nu\alpha}$, the equation simplifies to

$$g^{\nu\alpha}\left(\frac{1}{2} g_{\gamma\delta,\alpha} - g_{\gamma\alpha,\delta}\right)\dot{x}^{\gamma}\dot{x}^{\delta} - \ddot{x}^{\nu} = 0, \tag{4.83}$$

where we used $g^{\nu\alpha}g_{\gamma\alpha} = \delta^{\nu}_{\gamma}$ and $\delta^{\nu}_{\gamma}\ddot{x}^{\gamma} = \ddot{x}^{\nu}$. Recasting it to a form resembling Newton's laws, Eq. (4.83) becomes

$$\ddot{x}^{\nu} = -g^{\nu\alpha}\left(g_{\gamma\alpha,\delta} - \frac{1}{2} g_{\gamma\delta,\alpha}\right)\dot{x}^{\gamma}\dot{x}^{\delta}. \tag{4.84}$$

After recognizing

$$g_{\gamma\alpha,\delta}\dot{x}^{\gamma}\dot{x}^{\delta} = g_{\alpha\delta,\gamma}\dot{x}^{\gamma}\dot{x}^{\delta} \tag{4.85}$$

and rewriting

$$g_{\gamma\alpha,\delta}\dot{x}^{\gamma}\dot{x}^{\delta} = \frac{1}{2} g_{\gamma\alpha,\delta}\dot{x}^{\gamma}\dot{x}^{\delta} + \frac{1}{2} g_{\alpha\delta,\gamma}\dot{x}^{\gamma}\dot{x}^{\delta}, \tag{4.86}$$

we can express the geodesic equation in terms of the Christoffel symbols $\Gamma^{\nu}_{\gamma\delta}$, defined in Eq. (4.68).

The *geodesic equation* is

$$\ddot{x}^{\nu} = -\Gamma^{\nu}_{\gamma\delta}\dot{x}^{\gamma}\dot{x}^{\delta}. \tag{4.87}$$

It can be rewritten using the covariant derivative from Eq. (4.70):

$$\frac{D\dot{x}^{\nu}}{D\lambda} = \frac{d\dot{x}^{\nu}}{d\lambda} + \Gamma^{\nu}_{\gamma\delta}\dot{x}^{\gamma}\dot{x}^{\delta} = 0. \tag{4.88}$$

The geodesic equation above is a curved-spacetime analog of Newton's First Law $d\mathbf{u}/dt = 0$.

In the next section we show how the geodesic equation is really a curved-spacetime generalization of both Newton's First and Second Laws; the force as the gradient of the conservative potential in Newton's Second Law is represented by the spacetime metric elements.

In Euclidean space and Minkowski spacetime, $g_{\alpha\beta}$ is diagonal and constant, so its derivatives, and consequently the Christoffel symbols, vanish, thus leaving us with straight lines. In the flat space(time) limit, we thereby recover the results of Newtonian mechanics and special relativity, as we should.

Another advantage for using the Lagrangian in the form given in Eq. (4.80) is that solving the Lagrange equation in Eq. (4.79) in each coordinate yields the differential equation of the same form as the geodesic equation in Eq. (4.87). The Christoffel symbols can then simply be read off (see Appendix A).

4.9.1 Recovering Newtonian gravity

Let us verify that in the limit of slow motion ($v \ll c$) and weak, stationary gravitational fields, the geodesic equation yields Newton's Second Law.

In the limit of slow motion, the general expression of Eq. (4.87) reduces to $\Gamma^{\nu}_{00}(\dot{x}^0)^2$. However,

$$\Gamma^{\nu}_{00} = \frac{1}{2}g^{\nu\alpha}\left(g_{0\alpha,0} + g_{\alpha 0,0} - g_{00,\alpha}\right) = -\frac{1}{2}g^{\nu\alpha}g_{00,\alpha} = -\frac{1}{2}g^{\nu\beta}g_{00,\beta}, \tag{4.89}$$

as the stationary-field approximation make all time derivatives vanish, so $g_{\alpha\beta,0} = 0$. Using perturbation theory, we can recast the metric as a small deviation from a Minkowski flat spacetime:

$$g_{\alpha\beta} = \eta_{\alpha\beta} + \epsilon_{\alpha\beta}, \qquad g^{\alpha\beta} = \eta^{\alpha\beta} - \epsilon^{\alpha\beta}, \tag{4.90}$$

where $\epsilon_{\alpha\beta}$ is a small perturbation. Then, to the first order in $\epsilon_{\alpha\beta}$,

$$\Gamma^{\nu}_{00} = -\frac{1}{2}\left(\eta^{\nu\alpha} - \epsilon^{\nu\alpha}\right)\epsilon_{00,\alpha} = -\frac{1}{2}\eta^{\nu\alpha}\epsilon_{00,\alpha} + \mathcal{O}(\epsilon^2). \tag{4.91}$$

Then, $\Gamma^{0}_{00} = 0$ and $\Gamma^{\beta}_{00} = -\eta^{\beta\alpha}\epsilon_{00,\alpha}/2$. For $\nu = 0$, $\ddot{x}^0 = d^2(ct)/d^2\lambda = 0$ and $d(ct)/d\lambda =$ *const.*, and for $\nu = \beta$,

$$\ddot{x}^{\beta} = \frac{d^2x^{\beta}}{d^2\lambda} = \frac{1}{2}\eta^{\beta\alpha}\epsilon_{00,\alpha}(\dot{x}^0)^2 = \frac{1}{2}\eta^{\beta\alpha}\epsilon_{00,\alpha}\left(\frac{d(ct)}{d\lambda}\right)^2. \tag{4.92}$$

However,

$$\frac{dx^{\beta}}{d\lambda} = \frac{d(ct)}{d\lambda}\frac{dx^{\beta}}{d(ct)}, \tag{4.93}$$

hence,

$$\ddot{x}^{\beta} = \frac{d^2x^{\beta}}{d\lambda^2} = \left(\frac{d(ct)}{d\lambda}\right)^2\frac{d^2x^{\beta}}{d(ct)^2} \quad\Longrightarrow\quad \frac{d^2x^{\beta}}{d(ct)^2} = \frac{1}{2}\eta^{\beta\alpha}\epsilon_{00,\alpha}. \tag{4.94}$$

Recalling that in Newtonian mechanics the Euclidean 3-vector is $x^i = \left(x, y, z\right)$, and casting it in vector format we arrive at

$$\frac{d^2\boldsymbol{x}}{dt^2} = \frac{1}{2}c^2\nabla\epsilon_{00}. \tag{4.95}$$

Here, we used $\epsilon_{00,\alpha} = \nabla\epsilon_{00}$ in Euclidean space. When we compare this to Newton's Second Law,

$$\frac{d^2\boldsymbol{x}}{dt^2} = -\nabla\Phi, \tag{4.96}$$

we find that $\epsilon_{00} = -2\Phi/c^2$ and

$$g_{00} = -\left(1 + \frac{2\Phi}{c^2}\right). \tag{4.97}$$

In spherical symmetry, $\Phi = -GM/r$, hence,

$$g_{00} = -\left(1 - \frac{2GM}{c^2r}\right). \tag{4.98}$$

This quantifies how mass M curves the spacetime. Therefore Newtonian dynamics can be expressed in geometric form with a metric given by

$$ds^2 = -\left(1 + \frac{2\Phi}{c^2}\right)c^2dt^2 + dx^2 + dy^2 + dz^2. \tag{4.99}$$

Example 7: Find the geodesic in flat 2D space, as described by polar coordinates x^A, $A = 1, 2$, $x^1 = r$, and $x^2 = \phi$.
Here, the line element ds^2 satisfies

$$ds^2 = dr^2 + r^2 d\phi^2, \tag{4.100}$$

so that the metric is

$$g_{AB} = \begin{pmatrix} 1 & 0 \\ 0 & r^2 \end{pmatrix} \tag{4.101}$$

and the inverse metric is

$$g^{AB} = \begin{pmatrix} 1 & 0 \\ 0 & r^{-2} \end{pmatrix}. \tag{4.102}$$

By virtue of the definition,

$$\Gamma^A_{BC} = \frac{1}{2} g^{AD} \left(\frac{\partial g_{DB}}{\partial x^C} + \frac{\partial g_{DC}}{\partial x^B} - \frac{\partial g_{BC}}{\partial x^D} \right), \tag{4.103}$$

it follows that the only non-vanishing Christoffel symbols are

$$\Gamma^1_{22} = -r, \qquad \Gamma^2_{12} = \Gamma^2_{21} = \frac{1}{r}. \tag{4.104}$$

The two components of the geodesic equation,

$$\frac{d^2 x^A}{ds^2} = -\Gamma^A_{BC} \frac{dx^B}{ds} \frac{dx^C}{ds}, \tag{4.105}$$

thus become

$$\frac{d^2 r}{ds^2} = r \left(\frac{d\phi}{ds} \right)^2 \quad \text{and} \quad \frac{d^2 \phi}{ds^2} = -\frac{2}{r} \frac{d\phi}{ds} \frac{dr}{ds}. \tag{4.106}$$

The expression for the line element ds^2, given in Eq. (4.100) implies the first integral of motion,

$$\left(\frac{dr}{ds} \right)^2 + r^2 \left(\frac{d\phi}{ds} \right)^2 = const. \tag{4.107}$$

The equation for $d^2\phi/ds^2$ can be rewritten as

$$\frac{1}{r^2} \frac{d}{ds} \left(r^2 \frac{d\phi}{ds} \right) = 0, \tag{4.108}$$

which implies that

$$\frac{1}{r^2}\frac{d\phi}{ds} = \ell = const., \tag{4.109}$$

where ℓ is the second integral of motion. Inserting this result into the equation for $(dr/ds)^2$ yields

$$\frac{dr}{ds} = \left(1 - \frac{\ell}{r^2}\right)^{1/2}. \tag{4.110}$$

However, by combining this relation with the expression for $d\phi/ds$, one finds

$$\frac{d\phi}{ds} = \frac{d\phi/ds}{dr/ds} = \frac{\ell}{r^2}\left(1 - \frac{\ell}{r^2}\right)^{1/2}. \tag{4.111}$$

This expression can be integrated to yield

$$\phi = \phi_0 + \cos^{-1}\left(\frac{\ell}{r}\right), \tag{4.112}$$

where ϕ_0 is a constant of integration, so that

$$t\cos(\phi - \phi_0) = \ell. \tag{4.113}$$

Expanding out $\cos(\phi - \phi_0)$ in the usual fashion and noting that $x = r\cos\phi$ and $y = r\sin\phi$, one then concludes that

$$x\cos\phi_0 + y\sin\phi_0 = \ell, \tag{4.114}$$

i.e., that the geodesic equation corresponds to a straight line.

Exercises

1. Demonstrate explicitly that the quantity $\mathbf{A}\cdot\mathbf{B} = A_\alpha B^\alpha$ transforms as a scalar, *i.e.*, it is invariant under Lorentz transformation.
2. Find the geodesic equation in cylindrical coordinates.
3. Consider polar coordinates $ds^2 = dr^2 + r^2 d\theta^2$. Find the covariant derivative $V^\alpha_{;\alpha}$ of $V = r^2\cos\theta\,\boldsymbol{e}_r - \sin\theta\,\boldsymbol{e}_\theta$.
4. Using the definition of the covariant derivative in Eq. (4.69), show that
$$A^\gamma_{;\alpha\beta} - A^\gamma_{;\beta\alpha} = -R^\gamma_{\delta\alpha\beta}A^\delta.$$
5. Starting with the definition of the Christoffel symbols in Eq. (4.68), show that $\Gamma^\nu_{\alpha\beta} = \Gamma^\nu_{\beta\alpha}$.
6. Compute the components of the Riemann tensor on a surface of a unit sphere embedded in 3D space, where $ds^2 = d\theta^2 + \sin^2\theta\, d\phi^2$.

CHAPTER 5

Einstein's field equations

God used beautiful mathematics in creating the world.

Paul Dirac

The Big Picture: In the previous chapter, we derived the geodesic equation (a general relativity equivalent of Newton's First and Second Laws), which describes how a particle moves in spacetime. In this chapter, we are going to derive the second part necessary to complete the dynamical description: how the presence of matter and energy curves the ambient spacetime. This is given by Einstein's field equations, which are simply the general relativistic analog of the Poisson equation.

> **Note**: Oftentimes in the literature, "natural" units $c = 1$, $G = 1$, $h = 1$ are adopted, which simplify the resulting formulas and derivations. Obviously, it is easier to take derivatives with respect to t instead of ct. While natural units lead to compact and economical notation, there is value in using dimensionality as a "sanity check" for the derivations. For example, one has to make sure that only quantities that have the same units can be added together, or that only dimensionless quantities can be arguments in trigonometric, exponential, and other functions.

5.1. Bulk macroscopic description of matter: the Eulerian form

When describing the universe on the largest scales, as in astronomy or cosmology, it is useful to adopt a macroscopic approach that averages over the behavior of individual particles so as to identify bulk quantities like the mass density $\rho(\boldsymbol{x}, t)$, the pressure $P(\boldsymbol{x}, t)$, and the gravitational potential (scalar) field $\Phi(\boldsymbol{x}, t)$.

Questions related to bulk dynamics are then encapsulated through such relations as *conservation of mass* and *conservation of energy*, which say simply that matter and energy can be neither created nor destroyed, and *Euler's equation*, which is a macroscopic statement of $\boldsymbol{F} = m\boldsymbol{a}$. The equations expressing conservation of mass and $\boldsymbol{F} = m\boldsymbol{a}$ are easily derived in what follows.

A note of clarification is in order first: Euler's equations can be formulated in a *convective form* (also known as *Lagrangian form*) or in the *conservation form* (*Eulerian form*). The Lagrangian form quantifies the evolution of the properties of the system in a frame of reference that is moving with the particles. This is the form we are using in this book.

Relativity and Cosmology
https://doi.org/10.1016/B978-0-44-323542-9.00013-4

In contrast, the Eulerian form formulates equations of motion as conservation equations for a control volume fixed in space: for instance, the velocity (vector) field $\boldsymbol{v}(\boldsymbol{x}, t)$ for the particles is defined at a location $\boldsymbol{x}$. This form is used in fluid mechanics. For this section only, mostly for pedagogical reasons, we use the Eulerian form. Computation of bulk properties in the Lagrangian form will be discussed in Section 12.1.

Let the quantity

$$M = \int_V \rho(\boldsymbol{x}, t) d^3x \tag{5.1}$$

denote the total mass within some volume element V at time t. Conservation of mass then corresponds to the statement that if the density in some element d^3x changes over the course of time, this change must represent a net inward and outward flow of material. This means that

$$\frac{dM}{dt} = \int_V \frac{\partial \rho}{\partial t} d^3x = -\int_S \rho \boldsymbol{v} \cdot d^2\boldsymbol{S} = -\int_V \boldsymbol{\nabla} \cdot (\rho \boldsymbol{v}) d^3x, \tag{5.2}$$

where $\boldsymbol{\nabla} \equiv (\partial/\partial x^1, \partial/\partial x^2, \partial/\partial x^3)$. The final equality follows from the Divergence Theorem of calculus. Rearranging Eq. (5.2) yields

$$\int_V \left[\frac{\partial \rho}{\partial t} + \boldsymbol{\nabla} \cdot (\rho \boldsymbol{v}) \right] d^3x = 0. \tag{5.3}$$

Given, however, that this relation must hold for any volume V, one can conclude that the requirement that mass be conserved implies

$$\frac{\partial \rho}{\partial t} + \boldsymbol{\nabla} \cdot (\rho \boldsymbol{v}) = 0. \tag{5.4}$$

Suppose now that the fluid in question is characterized by an isotropic pressure, so that the distribution of velocities for the microscopic particles is the same in different directions. One can then use Newton's Second Law to write

$$\boldsymbol{F} = M \frac{d\boldsymbol{v}}{dt} = -\int_S P d^2\boldsymbol{S} - M \boldsymbol{\nabla} \Phi = -\int_V \boldsymbol{\nabla} P d^3x - M \boldsymbol{\nabla} \Phi, \tag{5.5}$$

where, again, we used the Divergence Theorem in the last step. The first term on the right-hand side of the equation above is the force due to internal pressure on the fluid element, and the second is due to its potential difference. Noting, however, that the mass M is the integral of the mass density, and assuming that the volume element V is small enough that $\boldsymbol{\nabla}\Phi$ and $d\boldsymbol{v}/dt$ can be considered constant (so that we can "slip" them inside the integral), we obtain

$$\int_V \rho \frac{d\boldsymbol{v}}{dt} d^3x = -\int_V [\boldsymbol{\nabla} P + \rho \boldsymbol{\nabla} \Phi] \, d^3x. \tag{5.6}$$

However, it is also apparent that $\boldsymbol{v} = \boldsymbol{v}(\boldsymbol{x}, t)$ will satisfy

$$\frac{d\boldsymbol{v}}{dt} = \frac{\partial \boldsymbol{v}}{\partial t} + \frac{\partial \boldsymbol{v}}{\partial \boldsymbol{x}}\frac{d\boldsymbol{x}}{dt} = \frac{\partial \boldsymbol{v}}{\partial t} + (\boldsymbol{v} \cdot \nabla)\boldsymbol{v}, \tag{5.7}$$

so that $\boldsymbol{F} = m\boldsymbol{a}$ reduces to Euler's equation in the form

$$\frac{\partial \boldsymbol{v}}{\partial t} + (\boldsymbol{v} \cdot \nabla)\boldsymbol{v} = -\frac{1}{\rho}\nabla P - \nabla \Phi. \tag{5.8}$$

5.1.1 Conservation laws in flat spacetime

In capturing an analog of conservation of mass or, equivalently, conservation of particle number in the context of special relativity in ordinary Cartesian coordinates, one needs to be careful to define things in a fashion that satisfies Lorentz invariance. As before, the coordinates of spacetime are (ct, x, y, z).

The quantity $dN = n^{(3)} d^3x$, the number of particles in d^3x, with the *number density* $n^{(3)}$ at time t, must be scalar, so that $dNdx^i$, defined in terms of an infinitesimal displacement vector dx^i, must transform as a vector. However, one can write:

$$dNdx^i = n^{(3)} d^3x dx^i = n^{(3)} \frac{dx^i}{dt} dt d^3x = n^{(3)} \frac{dx^i}{dt} d^4x. \tag{5.9}$$

The fact that d^4x is a Lorentz scalar thus implies that the quantity

$$n^{(3)} \frac{dx^i}{dt} = n^{(3)} \frac{d\tau}{dt}\frac{dx^i}{d\tau} \equiv nu^i \tag{5.10}$$

must transform as a four-vector. The quantity

$$n \equiv n^{(3)} \frac{d\tau}{dt} = n^{(3)} \sqrt{1 - v^2/c^2} \tag{5.11}$$

must transform as a scalar.

The 3D law expressing conservation of number

$$\frac{\partial n^{(3)}}{\partial t} + \frac{\partial}{\partial x^i}\left(n^{(3)} \frac{dx^i}{dt}\right) = 0, \tag{5.12}$$

can thus be rewritten in the manifestly covariant form

$$\frac{\partial}{\partial x^\alpha}(nu^\alpha) = 0. \tag{5.13}$$

This picture combines the number density scalar $n^{(3)}$ and the number current three-vector $n^{(3)} v^i$ of Newtonian physics into a single four-vector: the temporal component describes density and the spatial components describe flows. When endeavoring to capture the notion of conservation of energy and momentum, one is led to proceed in a similar fashion. In relativity, one has a description in which energy and the ordinary three-momentum are combined into a single energy–momentum four-vector. Therefore the appropriate analog of the number density four-vector must contain information about 16 quantities: the density of energy and three-momentum (4 quantities) and the flows of energy and three-momentum in three different spatial directions ($4 \times 3 = 12$ quantities). This can be achieved through the identification of a 4×4 matrix $T^{\alpha\beta}$ (formally, a contravariant second-rank tensor). The quantity $T^{\alpha\beta}$ is called an *energy–momentum tensor* and is described in the next section.

5.2. The energy–momentum tensor

The energy–momentum (stress–energy) tensor $T^{\alpha\beta}$ describes the density and the flows of the four-momentum $(-E/c, p^1, p^2, p^3)$. The component $T^{\alpha\beta}$ is the flux or flow of the α component of the four-momentum crossing the hypersurface of constant x^β (see Fig. 5.1):

- T^{00} represents energy density;
- $T^{0\alpha}$ represents the flow (flux) of energy in the x^α direction;
- $T^{\alpha 0}$ represents the density of the α-component of momentum;
- $T^{\alpha\beta}$ represents the flow of the α-component of momentum in the β-direction (stress).

The energy–momentum tensor is symmetric $T^{\alpha\beta} = T^{\beta\alpha}$. In general, the form of $T^{\alpha\beta}$ depends on the microscopic dynamics of the material that one is modeling. We now consider two types of momentum–energy tensor frequently used in general relativity: perfect fluid and dust.

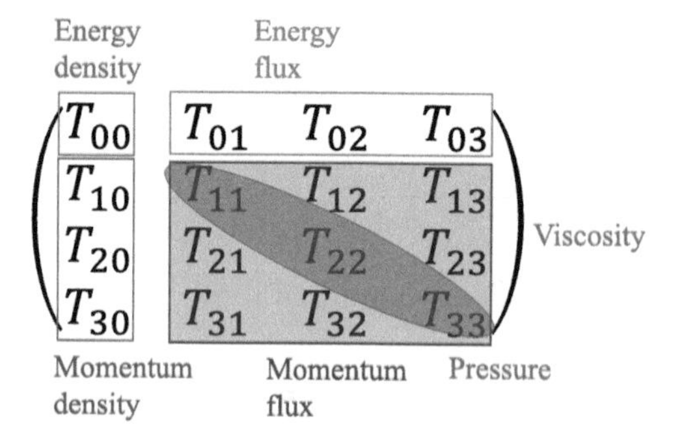

Figure 5.1 Components of the energy–momentum tensor $T_{\alpha\beta}$.

Perfect fluid is defined by the requirement that, in its rest frame:

- T^{00} is the locally measured energy density;
- $T^{0\alpha} = T^{\alpha 0} = 0$, so that there is no net flow of energy and no net three-momentum density;
- The flow of particles is the same in all directions (the pressure is isotropic: $T^{ij} = P\delta^{ij}$), which means that $T^{\alpha\beta}$ is diagonal (no net shears).

The perfect fluid has no heat conduction or viscosity. It is fully parametrized by its mass density ρ and the pressure P. It is given by

$$T^{\alpha\beta} = \left(\rho + \frac{P}{c^2}\right) u^\alpha u^\beta + P g^{\alpha\beta}. \tag{5.14}$$

Both terms ρc^2 and P represent physical energy density. For a *comoving* observer – for whom the spatial coordinates (r, θ, ϕ) remain fixed, while the entire space may scale due to an evolving Universe, discussed later in the text – the four-velocity is given by $\mathbf{u} = (c, 0, 0, 0)$, so the stress–energy tensor reduces to

$$T^{\alpha\beta} = \begin{pmatrix} \rho c^2 & 0 & 0 & 0 \\ 0 & P & 0 & 0 \\ 0 & 0 & P & 0 \\ 0 & 0 & 0 & P \end{pmatrix}. \tag{5.15}$$

In the limit of $P \to 0$, the perfect-fluid approximation reduces to that of dust. Perfect fluid is an approximation of the Universe at earlier times, when radiation dominates.

Dust is the simplest possible energy–momentum tensor (essentially perfect fluid with $P = 0$). It is given by

$$T^{\alpha\beta} = \rho u^\alpha u^\beta. \tag{5.16}$$

For a comoving observer the four-velocity is given by $\mathbf{u} = (c, 0, 0, 0)$, so the stress–energy tensor reduces to

$$T^{\alpha\beta} = \begin{pmatrix} \rho c^2 & 0 & 0 & 0 \\ 0 & 0 & 0 & 0 \\ 0 & 0 & 0 & 0 \\ 0 & 0 & 0 & 0 \end{pmatrix}. \tag{5.17}$$

Dust is an approximation of the Universe at later times, when matter component dominates and radiation (pressure) is negligible.

Conservation equations for the energy–momentum tensor $T^{\alpha\beta}$ are simply given by

$$T^{\alpha\beta}_{;\beta} = 0. \tag{5.18}$$

This expression incorporates both energy and momentum conservations in a general metric. In the limit of flat spacetime (Minkowski metric), according to the Principle of General Covariance (see Eq. (4.71)), it reduces to

$$T^{\alpha\beta}_{,\beta} \equiv \frac{\partial T^{\alpha\beta}}{\partial x^{\beta}} = 0, \tag{5.19}$$

from which the traditional expressions for the conservation of momentum and energy are readily recovered.

5.3. The Riemann tensor, the Ricci tensor, the Ricci scalar, and the Einstein tensor

The Riemann (curvature) tensor [8] plays an important role in specifying the geometrical properties of spacetime. It is defined in terms of Christoffel symbols:

$$R^{\alpha}_{\beta\gamma\delta} \equiv \Gamma^{\alpha}_{\beta\delta,\gamma} - \Gamma^{\alpha}_{\beta\gamma,\delta} + \Gamma^{\nu}_{\beta\delta}\Gamma^{\alpha}_{\nu\gamma} - \Gamma^{\nu}_{\beta\gamma}\Gamma^{\alpha}_{\nu\delta}, \tag{5.20}$$

where $\Gamma^{\alpha}_{\beta\delta,\gamma} \equiv \partial\Gamma^{\alpha}_{\beta\delta}/\partial x^{\gamma}$. The spacetime is considered flat if the Riemann tensor vanishes everywhere.

Using the definition of the Christoffel symbols

$$\Gamma^{\nu}_{\alpha\beta} \equiv \frac{1}{2} g^{\nu\gamma}\left(g_{\alpha\gamma,\beta} + g_{\gamma\beta,\alpha} - g_{\alpha\beta,\gamma}\right), \tag{5.21}$$

the contravariant index in the Riemann tensor above can be lowered to yield (after recalling $g_{\mu\nu}g^{\nu\gamma} = \delta^{\gamma}_{\mu}$):

$$R_{\alpha\beta\gamma\delta} \equiv \frac{1}{2}\left(g_{\beta\gamma,\alpha\delta} + g_{\alpha\delta,\beta\gamma} - g_{\beta\delta,\alpha\gamma} - g_{\alpha\gamma,\beta\delta}\right) + g_{\alpha\mu}\Gamma^{\nu}_{\beta\delta}\Gamma^{\mu}_{\nu\gamma} - g_{\alpha\mu}\Gamma^{\nu}_{\beta\gamma}\Gamma^{\mu}_{\nu\delta}. \tag{5.22}$$

$R_{\alpha\beta\gamma\delta}$ corresponds to a $4 \times 4 \times 4 \times 4$ tensor, so it has $4^4 = 256$ numbers.

Eq. (5.22) reveals the properties of the Riemann tensor:

- *Skew symmetry* (changing order of indices within the same pair flips the sign of the value of the tensor):

$$R_{\alpha\beta\gamma\delta} = -R_{\beta\alpha\gamma\delta} = -R_{\alpha\beta\delta\gamma}. \tag{5.23}$$

- *Interchange symmetry* (interchanging pairs of indices does not change the value of the tensor):

$$R_{\alpha\beta\gamma\delta} = R_{\gamma\delta\alpha\beta}. \tag{5.24}$$

- *First Bianchi identity*:

$$R_{\alpha\beta\gamma\delta} + R_{\alpha\gamma\delta\beta} + R_{\alpha\delta\beta\gamma} = 0. \tag{5.25}$$

Due to the symmetries above, the Riemann tensor in 4D spacetime has only 20 independent components. The general rule for computing the number of independent components in an N-dimensional spacetime is $N^2(N^2 - 1)/12$.

As the Riemann tensor involves derivatives of the metric tensor, in the special case of a flat spacetime $R_{\alpha\beta\gamma\delta} = 0$.

The Ricci tensor is obtained from the Riemann tensor by simply contracting over two of the indices:

$$R_{\alpha\beta} \equiv R^{\gamma}_{\ \alpha\gamma\beta} \equiv g^{\delta\gamma} R_{\delta\alpha\gamma\beta}. \tag{5.26}$$

It is symmetric, which means that it has at most 10 independent quantities.

The Ricci scalar is obtained by contracting the Ricci tensor over the remaining two indices:

$$\mathcal{R} \equiv g^{\alpha\beta} R_{\alpha\beta} = R^{\alpha}_{\alpha}. \tag{5.27}$$

The Bianchi identities are another important symmetry of the Riemann tensor,

$$R_{\alpha\beta\gamma\delta;\nu} + R_{\alpha\beta\delta\nu;\gamma} + R_{\alpha\beta\nu\gamma;\delta} = 0, \tag{5.28}$$

which, after contracting, leads to

$$R^{\alpha\beta}_{\ \ ;\alpha} = \frac{1}{2} g^{\alpha\beta} \mathcal{R}_{;\alpha}. \tag{5.29}$$

The Einstein tensor is defined in terms of the Ricci tensor and Ricci scalar as

$$G_{\alpha\beta} \equiv R_{\alpha\beta} - \frac{1}{2} g_{\alpha\beta} R. \tag{5.30}$$

From Eq. (5.29), a very important property of the Einstein tensor is derived:

$$G_{\alpha\beta;\alpha} = 0. \tag{5.31}$$

5.4. Evolution of energy density

Conservation of energy given in Eq. (5.18) can be used to determine how components of the energy–momentum tensor evolve with time. The mixed energy–momentum tensor for perfect fluid is

$$T^{\alpha}_{\beta} = \begin{pmatrix} -\rho c^2 & 0 & 0 & 0 \\ 0 & P & 0 & 0 \\ 0 & 0 & P & 0 \\ 0 & 0 & 0 & P \end{pmatrix} \tag{5.32}$$

and its conservation is given by a covariant generalization of Eq. (5.18):

$$T^{\mu}_{\nu;\mu} \equiv \frac{\partial T^{\mu}_{\nu}}{\partial x^{\mu}} + \Gamma^{\mu}_{\alpha\mu} T^{\alpha}_{\nu} - \Gamma^{\alpha}_{\nu\mu} T^{\mu}_{\alpha} = 0, \tag{5.33}$$

which gives four separate equations. Consider the $\nu = 0$ component:

$$\frac{\partial T^{\mu}_{0}}{\partial x^{\mu}} + \Gamma^{\mu}_{\alpha\mu} T^{\alpha}_{0} - \Gamma^{\alpha}_{0\mu} T^{\mu}_{\alpha} = 0. \tag{5.34}$$

Due to isotropy, all non-diagonal terms of T^{α}_{β} vanish, hence, $T^{\alpha}_{0} \neq 0$ only for $\alpha = 0$. This leads to $\mu = 0$ in the first term and $\alpha = 0$ in the second term above. Setting $\mu = 0$ in Eq. (5.34), so that $x^0 = ct$, and using $T^0_0 = -\rho c^2$, we obtain

$$\begin{aligned} -\frac{\partial \rho c^2}{\partial ct} - \Gamma^{\mu}_{0\mu} \rho c^2 - \Gamma^{\alpha}_{0\mu} T^{\mu}_{\alpha} &= 0, \\ \dot{\rho} c + \Gamma^{\mu}_{0\mu} \rho c^2 + \Gamma^{\alpha}_{0\mu} T^{\mu}_{\alpha} &= 0, \end{aligned} \tag{5.35}$$

where $\dot{} \equiv \partial/\partial t$. Expanding flat spacetime is described by the flat Friedmann–Lemaître–Robertson–Walker metric tensor given in Eq. (4.27):

$$g_{\alpha\beta} = \begin{pmatrix} -1 & 0 & 0 & 0 \\ 0 & a^2(t) & 0 & 0 \\ 0 & 0 & a^2(t) & 0 \\ 0 & 0 & 0 & a^2(t) \end{pmatrix} \tag{5.36}$$

and, of course, its contravariant analog:

$$g^{\alpha\beta} = \begin{pmatrix} -1 & 0 & 0 & 0 \\ 0 & a^{-2}(t) & 0 & 0 \\ 0 & 0 & a^{-2}(t) & 0 \\ 0 & 0 & 0 & a^{-2}(t) \end{pmatrix}. \tag{5.37}$$

From the definition of the Christoffel symbols,

$$\Gamma^{\alpha}_{\nu\mu} \equiv \frac{1}{2} g^{\alpha\gamma} \left(g_{\nu\gamma,\mu} + g_{\gamma\mu,\nu} - g_{\nu\beta,\gamma} \right), \tag{5.38}$$

$$\begin{aligned} \Gamma^{\alpha}_{0\mu} &= \frac{1}{2} g^{\alpha\gamma} \left(g_{0\gamma,\mu} + g_{\gamma\mu,0} - g_{0\beta,\gamma} \right) = \frac{1}{2} g^{\alpha\gamma} g_{\gamma\mu,0}, \qquad \text{because } g_{0\gamma} = const.,\ g_{0\beta} = const., \\ &= \begin{cases} \frac{1}{2} \left(\delta_{\alpha\gamma} a^{-2} \right) \left(2\delta_{\gamma\mu} \dot{a} a / c \right) & \text{if } \alpha \neq 0 \text{ and } \mu \neq 0, \\ 0 & \text{if } \alpha = 0 \text{ or } \mu = 0, \end{cases} \qquad \text{because } g_{\gamma 0,0} = g_{0\mu,0} = 0;\ \frac{\partial}{\partial ct} = \frac{1}{c}\frac{\partial}{\partial t}, \\ &= \begin{cases} \dot{a}/(ca)\delta_{ij} & i,j = 1,2,3, \\ 0 & \text{otherwise}, \end{cases} \end{aligned} \tag{5.39}$$

so that the only non-zero $\Gamma^{\alpha}_{0\mu}$ is $\Gamma^{1}_{01} = \Gamma^{2}_{02} = \Gamma^{3}_{03} = \dot{a}/(ca)$ (note: when summed over repeated indices $\Gamma^{i}_{0i} = 3\dot{a}/(ca)$). Hence, the conservation law in the expanding Universe from Eq. (5.35) becomes, after using $\delta_{ij} T^{j}_{i} = T_{ii} = 3P$:

$$\dot{\rho} c + 3\frac{\dot{a}}{a}\rho c + \frac{\dot{a}}{ca} 3P = 0 \quad \rightarrow \quad \dot{\rho} + 3\left(\rho + \frac{P}{c^2}\right)\frac{\dot{a}}{a} = 0. \tag{5.40}$$

We can massage this to obtain

$$\dot{\rho} + 3\rho\frac{\dot{a}}{a} = -3\frac{\dot{a}}{a}\frac{P}{c^2} \quad \rightarrow \quad a^{-3}\frac{\partial}{\partial t}\left(\rho a^3\right) = -3\frac{\dot{a}}{a}\frac{P}{c^2} \quad \rightarrow \quad \frac{\partial}{\partial t}\left(\rho a^3\right) = -3a^2\dot{a}\frac{P}{c^2} \tag{5.41}$$

and use it to find out how both matter and radiation scale with expansion. Note that the ρ is the volumetric *mass* density of matter, so that its corresponding volumetric *energy* density is given via $E = mc^2$ as $u = \rho c^2$.

5.4.1 Matter density

For non-relativistic matter (dust approximation), we have zero pressure $P_{\mathrm{m}} = 0$, so Eq. (5.41) yields

$$\frac{\partial}{\partial t}\left(\rho_{\mathrm{m}} a^3\right) = -3a^2\dot{a}\frac{P_{\mathrm{m}}}{c^2} = 0, \tag{5.42}$$

which means that $\rho_{\mathrm{m}} \propto a^{-3}$. This should come as no surprise: the total mass M_{m} is conserved, and the volume of the Universe goes as $V \propto a^3$, so $\rho_{\mathrm{m}} \propto M_{\mathrm{m}}/V \propto a^{-3}$.

5.4.2 Radiation density

For radiation (perfect-fluid approximation), $P_{\mathrm{r}} = u/3$ (see the derivation in Appendix J), where u is the radiation energy density (photons and neutrinos). A direct way of deriving how the radiation energy density scales with the scale factor a is to substitute $u = \rho c^2$ in Eq. (5.40). However, what we choose to do instead is to treat matter and radiation content on an equal footing, which would have come out easily if we chose to follow natural units, where $c = 1$, and $E = mc^2$ simply becomes $E = m$. That essentially erases the distinction between energy and mass density. Here, we will track the mass density equivalent ρ_{r} of the radiation energy density u_{r}, which are simply related as $u_{\mathrm{r}} = \rho_{\mathrm{r}} c^2$. Then, $P_{\mathrm{r}} = \rho_{\mathrm{r}} c^2$ and we can use Eq. (5.40) to obtain

$$\dot{\rho}_{\mathrm{r}} + 3\left(\rho_{\mathrm{r}} + P_{\mathrm{r}}\right)\frac{\dot{a}}{a} = 0 \quad \rightarrow \quad \dot{\rho}_{\mathrm{r}} + 3\left(\rho_{\mathrm{r}} + \frac{\rho_{\mathrm{r}}}{3}\right)\frac{\dot{a}}{a} = 0 \tag{5.43}$$

$$\rightarrow \quad \dot{\rho}_{\mathrm{r}} + 4\rho_{\mathrm{r}}\frac{\dot{a}}{a} = a^{-4}\frac{\partial}{\partial t}\left(\rho_{\mathrm{r}} a^4\right) = 0 \quad \rightarrow \quad \rho_{\mathrm{r}} a^4 = const., \tag{5.44}$$

which implies that $\rho_{\mathrm{r}} \propto a^{-4}$. This too should not surprise us – since radiation density is directly proportional to the energy per particle ($E_1 = 2\pi\hbar\nu$, where $\nu \propto 1/\lambda$ is the frequency and λ is the wavelength; the total number of radiation particles N_{r} is assumed constant) and inversely proportional to the total volume (which, as seen above $V \propto a^{-3}$), *i.e.*,

$$\rho_{\mathrm{r}} \propto \frac{N_{\mathrm{r}} 2\pi\hbar\nu}{V} \propto \frac{N_{\mathrm{r}} 2\pi\hbar}{\lambda V} \propto a^{-4}, \tag{5.45}$$

as $\lambda \propto a$. The last part states that the energy per particle decreases as the Universe expands.

5.5. Einstein's field equations

The stage is now set for deriving and understanding Einstein's field equations [9,10].

General relativity must present appropriate analogs of the two parts of the Newtonian dynamical picture: (1) how particles move in response to gravity; and (2) how particles generate gravitational effects. The first part was answered when we derived the geodesic equation as the analog of Newton's First and Second Laws. The second part requires finding the analog of the Poisson equation,

$$\nabla^2 \Phi(\mathbf{x}) = 4\pi G \rho(\mathbf{x}), \tag{5.46}$$

which specifies how matter curves spacetime. All equations in general relativity must be in tensor form. Arguably the most enlightening derivation of the Einstein's equations is to argue about its form on physical grounds, which was the approach originally adopted by Einstein.

In Newtonian gravity, the rest mass generates gravitational effects. From special relativity, however, we learned that the rest mass is just one form of energy, and that the mass and energy are equivalent. Therefore we should expect that in general relativity all sources of both energy and momentum contribute to generating spacetime curvature (one observer's momentum may be another's energy). In general relativity, the energy–momentum tensor $T^{\alpha\beta}$ is the source for spacetime curvature in the same sense that the mass density ρ is the source for the potential Φ. Hence, at this point, we can say that we have a pretty good guess of what the right-hand side of the general relativistic analog of the Poisson equation should be: $\kappa T^{\alpha\beta}$ (with κ some constant to be determined later).

What about the left-hand side of the general relativistic analog of the Poisson equation? What is analogous to $\nabla^2 \Phi(\mathbf{x})$? As we have seen earlier (Eq. (4.97)), the spacetime metric in the Newtonian limit is modified by a term proportional to Φ. If we extend this analogy, then the general-relativistic counterpart of $\nabla \Phi$ in the right-hand side of Newton's Second Law should include derivatives of the metric, which is indeed verified by the form of the geodesic equation (see Eqs. (4.68) and (4.87)). Further extending this analogy, one would expect that the general-relativistic counterpart of $\nabla^2 \Phi(\mathbf{x})$ would contain terms that involve second derivatives of the metric. From Eq. (5.22), we see that the Riemann tensor $R_{\alpha\beta\gamma\delta}$ – and consequently its contractions Ricci tensor $R_{\alpha\beta}$ and Ricci scalar $\mathcal{R}$ – contain second derivatives of the metric, and thus become viable candidates for the left-hand side of Einstein's field equation.

Led by this line of reasoning, Einstein originally suggested that the field equation might read:

$$R_{\alpha\beta} = \kappa T_{\alpha\beta}, \tag{5.47}$$

but it was quickly recognized (with a little help from Hilbert [11]) that this *cannot* be correct, because while the conservation of energy–momentum requires $T^{\alpha\beta}_{\ ;\alpha} = 0$, the same is in general not true of the Ricci tensor: $R^{\alpha\beta}_{\ ;\alpha} \neq 0$. Fortunately, Einstein's tensor $G_{\alpha\beta}$ (a combination of the Ricci tensor and the Ricci scalar), satisfies the requirement that it has vanishing divergence. Therefore Einstein's equation then becomes

$$G_{\alpha\beta} \equiv R_{\alpha\beta} - \frac{1}{2} g_{\alpha\beta} \mathcal{R} = \kappa T_{\alpha\beta}, \tag{5.48}$$

where $G_{\alpha\beta}$ is the *Einstein tensor*. By matching Einstein's equation in the Newtonian limit to the Poisson equation, the constant κ is found to be $8\pi G/c^4$.

Finally, *Einstein's field equations* are:

$$G_{\alpha\beta} = R_{\alpha\beta} - \frac{1}{2}g_{\alpha\beta}\mathcal{R} = \frac{8\pi G}{c^4}T_{\alpha\beta}. \tag{5.49}$$

Exercises

1. Use Eq. (5.20) to express the Ricci tensor in terms of Christoffel symbols.
2. Derive Eq. (5.22) from Eq. (5.20) and Eq. (5.21).
3. Derive Eq. (5.25).
4. Derive Eq. (5.31) from Eqs. (5.29) and (5.30).
5. Compute the Ricci scalar $\mathcal{R}$ on the surface of a unit sphere embedded in 3D space. (You can use the results from Exercise 6 in Chapter 4.)

CHAPTER 6

The Schwarzschild metric and black holes

All of physics is either impossible or trivial. It is impossible until you understand it, and then it becomes trivial.

Ernest Rutherford

The Big Picture: In the previous chapter, we derived Einstein's field equations, which govern how the presence of matter and energy curves the ambient spacetime. Just a few months after Einstein published his field equations in 1915, their first solution was derived by Schwarzschild for the spacetime curvature near an isolated massive object. In the limit of the object being not so massive and compact, the Newtonian description is recovered. The other limit, that of very massive and compact objects, reveals the existence of a spacetime singularity, which we now know as "black holes."

6.1. The Schwarzschild problem

Shortly after Einstein published his field equations of general relativity, Karl Schwarzschild [12] solved them to find the spacetime metric outside a single, isolated, stationary, neutrally charged,[1] non-rotating,[2] spherical distribution of matter of mass M. Since the space outside the distribution is empty, the energy–momentum tensor $T_{\alpha\beta}$ vanishes, so Einstein's field equations become

$$R_{\alpha\beta} - \frac{1}{2} g_{\alpha\beta} \mathcal{R} = 0, \tag{6.1}$$

with an appropriate metric tensor.

The appropriate boundary conditions for the Schwarzschild problem – solutions of Einstein's equations outside a single, isolated, stationary spherical distribution of matter – are:

1. The metric must match the interior metric at the body's surface;
2. The metric must go to the flat (Minkowski) metric far away from the body.

[1] The solutions for the non-neutral (non-zero net charge) isolated mass are known as the Reissner–Nordström metrics.

[2] The solutions for the rotating isolated mass are known as the Kerr metrics [13].

Relativity and Cosmology
https://doi.org/10.1016/B978-0-44-323542-9.00014-6

We now derive the Schwarzschild metric $g_{\alpha\beta}$ that solves the Schwarzschild problem.

A general static and isotropic metric satisfies:
1. **Static:** *both* time independent *and* symmetric under time reversal (*only* time independent $\leftrightarrow$ *stationary*);
2. **Isotropic:** invariant under spatial rotations (same in all directions).

The interval satisfying these criteria may be written as

$$ds^2 = -A(r)c^2dt^2 + B(r)dr^2 + r^2d\theta^2 + r^2\sin^2\theta\, d\phi^2, \tag{6.2}$$

where the first two terms on the right-hand side describe radial behavior (isotropy) and the last two the surface of the sphere (spherical symmetry). It can be expressed in many equivalent forms. One convenient form is:

$$ds^2 = -e^{N(r)}c^2dt^2 + e^{P(r)}dr^2 + r^2d\theta^2 + r^2\sin^2\theta\, d\phi^2, \tag{6.3}$$

corresponding to the metric tensor,

$$g_{\alpha\beta} = \begin{pmatrix} -e^{N(r)} & 0 & 0 & 0 \\ 0 & e^{P(r)} & 0 & 0 \\ 0 & 0 & r^2 & 0 \\ 0 & 0 & 0 & r^2\sin^2\theta \end{pmatrix}, \tag{6.4}$$

where the spacetime coordinates are $x^\alpha = (ct, r, \theta, \phi)$. The Schwarzschild problem reduces to solving for $N(r)$ and $P(r)$ from Einstein's field equations and the appropriate boundary conditions.

6.2. Solving the Schwarzschild problem

Earlier, in Eq. (4.80), we defined an alternative Lagrangian:

$$L = \frac{1}{2}g_{\alpha\beta}\dot{x}^\alpha\dot{x}^\beta, \tag{6.5}$$

where the dot denotes a λ-derivative: $\dot{x}^\alpha \equiv dx^\alpha/d\lambda$ that for the metric in Eq. (6.4) becomes:

$$L = -\frac{1}{2}e^N c^2\dot{t}^2 + \frac{1}{2}e^P\dot{r}^2 + \frac{1}{2}r^2\dot{\theta}^2 + \frac{1}{2}r^2\sin^2\theta\dot{\phi}^2. \tag{6.6}$$

This alternative Lagrangian allows us to easily read off the Christoffel symbols by comparing it to the geodesic equation in Eq. (4.87) (see Appendices A and B):

$$\ddot{x}^\nu + \Gamma^\nu_{\gamma\delta}\dot{x}^\gamma\dot{x}^\delta = 0, \tag{6.7}$$

which we can combine to obtain the Riemann and Ricci tensors. Let us solve the Lagrange equations

$$\frac{\partial L}{\partial x^\alpha} - \frac{d}{d\lambda}\frac{\partial L}{\partial \dot{x}^\alpha} = 0, \tag{6.8}$$

for each of the components of the spacetime. Here, $'$ will denote an r-derivative: $P'(r) = dP(r)/dr$. Note that, by the chain rule, $dP(r)/d\lambda = (dP(r)/dr)(dr/(d\lambda)) = P'(r)\dot{r}$. This is, of course, by virtue of $r = r(\lambda)$, and applies to other functions of r, such as $N(r)$.

- ct-component:

$$\begin{aligned}
\frac{\partial L}{\partial ct} - \frac{d}{d\lambda}\left(\frac{\partial L}{\partial c\dot{t}}\right) &= 0 \\
0 - \frac{d}{d\lambda}\left(-e^N c\dot{t}\right) &= 0 \\
e^N \frac{dN}{dr}\frac{dr}{d\lambda} c\dot{t} + e^N c\ddot{t} &= 0 \\
e^N\left(\ddot{t} + N'\dot{t}\dot{r}\right) &= 0 \\
\ddot{t} + N'\dot{t}\dot{r} &= 0.
\end{aligned} \tag{6.9}$$

  After comparing it to Eq. (6.7), we obtain

$$\ddot{t} + \left(\Gamma^0_{01} + \Gamma^0_{10}\right)\dot{t}\dot{r} = 0, \tag{6.10}$$

  which means that (because of the symmetry of the Christoffel symbols: $\Gamma^\alpha_{\beta\gamma} = \Gamma^\alpha_{\gamma\beta}$)

$$\Gamma^0_{01} = \Gamma^0_{10} = \frac{1}{2}N', \tag{6.11}$$

  while the remaining $\Gamma^0_{\alpha\beta}$ symbols vanish.

- r-component:

$$\begin{aligned}
\frac{\partial L}{\partial r} - \frac{d}{d\lambda}\left(\frac{\partial L}{\partial \dot{r}}\right) &= 0 \\
-\frac{1}{2}N'e^N\dot{t}^2 + \frac{1}{2}P'e^P\dot{r}^2 + r\dot{\theta}^2 + r\sin^2\theta\dot{\phi}^2 - \frac{d}{d\lambda}\left(e^P\dot{r}\right) &= 0 \\
-\frac{1}{2}N'e^N\dot{t}^2 + \frac{1}{2}P'e^P\dot{r}^2 + r\dot{\theta}^2 + r\sin^2\theta\dot{\phi}^2 - e^P P'\dot{r}^2 - e^P\ddot{r} &= 0 \\
-e^P\left(\ddot{r} + \frac{1}{2}N'e^{N-P}\dot{t}^2 + \frac{1}{2}P'\dot{r}^2 - e^{-P}r\dot{\theta}^2 - e^{-P}r\sin^2\theta\dot{\phi}^2\right) &= 0 \\
\ddot{r} + \frac{1}{2}N'e^{N-P}\dot{t}^2 + \frac{1}{2}P'\dot{r}^2 - e^{-P}r\dot{\theta}^2 - e^{-P}r\sin^2\theta\dot{\phi}^2 &= 0.
\end{aligned} \tag{6.12}$$

After comparing it to Eq. (6.7), we obtain

$$\ddot{r} + \Gamma^1_{00}\dot{t}^2 + \Gamma^1_{11}\dot{r}^2 + \Gamma^1_{22}\dot{\theta}^2 + \Gamma^1_{33}\dot{\phi}^2 = 0, \tag{6.13}$$

which means that

$$\begin{aligned}
\Gamma^1_{00} &= \frac{1}{2}N'e^{N-P}, \\
\Gamma^1_{11} &= \frac{1}{2}P', \\
\Gamma^1_{22} &= -e^{-P}r, \\
\Gamma^1_{33} &= -e^{-P}r\sin^2\theta,
\end{aligned} \tag{6.14}$$

while the remaining $\Gamma^1_{\alpha\beta}$ symbols vanish.

- θ-component:

$$\begin{aligned}
\frac{\partial L}{\partial \theta} - \frac{d}{d\lambda}\left(\frac{\partial L}{\partial \dot{\theta}}\right) &= 0 \\
\frac{1}{2}r^2 2\sin\theta\cos\theta\dot{\phi}^2 - \frac{d}{d\lambda}\left(r^2\dot{\theta}\right) &= 0 \\
\frac{1}{2}r^2\sin 2\theta\dot{\phi}^2 - 2r\dot{r}\dot{\theta} - r^2\ddot{\theta} &= 0 \\
-r^2\left(\ddot{\theta} - \frac{1}{2}\sin 2\theta\dot{\phi}^2 + 2\frac{\dot{r}}{r}\dot{\theta}\right) &= 0 \\
\ddot{\theta} + \frac{2}{r}\dot{r}\dot{\theta} - \frac{1}{2}\sin 2\theta\dot{\phi}^2 &= 0.
\end{aligned} \tag{6.15}$$

After comparing it to Eq. (6.7), we obtain

$$\ddot{\theta} + \left(\Gamma^2_{12} + \Gamma^2_{21}\right)\dot{r}\dot{\theta} + \Gamma^2_{33}\dot{\phi}^2 = 0, \tag{6.16}$$

which means that

$$\begin{aligned}
\Gamma^2_{12} = \Gamma^2_{21} &= \frac{1}{r}, \\
\Gamma^2_{33} &= -\frac{1}{2}\sin 2\theta,
\end{aligned} \tag{6.17}$$

while the remaining $\Gamma^2_{\alpha\beta}$ symbols vanish.

- ϕ-component:

$$\begin{aligned}
\frac{\partial L}{\partial \phi} - \frac{d}{d\lambda}\left(\frac{\partial L}{\partial \dot{\phi}}\right) &= 0 \\
0 - \frac{d}{d\lambda}\left(r^2\sin^2\theta\dot{\phi}\right) &= 0
\end{aligned}$$

$$-2r\dot{r}\sin^2\theta\dot{\phi} - 2r^2\sin\theta\cos\theta\dot{\theta}\dot{\phi} - r^2\sin^2\theta\ddot{\phi} = 0$$
$$-r^2\sin^2\theta\left(\ddot{\phi} + 2\frac{\dot{r}}{r}\dot{\phi} + 2\frac{\cos\theta}{\sin\theta}\dot{\theta}\dot{\phi}\right) = 0$$
$$\ddot{\phi} + \frac{2}{r}\dot{r}\dot{\phi} + 2\cot\theta\dot{\theta}\dot{\phi} = 0. \tag{6.18}$$

After comparing it to Eq. (6.7), we obtain

$$\ddot{\phi} + \left(\Gamma^3_{13} + \Gamma^3_{31}\right)\dot{r}\dot{\phi} + \left(\Gamma^3_{23} + \Gamma^3_{32}\right)\dot{\theta}\dot{\phi} = 0, \tag{6.19}$$

which means that

$$\Gamma^3_{13} = \Gamma^3_{31} = \frac{1}{r},$$
$$\Gamma^3_{23} = \Gamma^3_{32} = \cot\theta, \tag{6.20}$$

while the remaining $\Gamma^3_{\alpha\beta}$ symbols vanish.

The Christoffel symbols associated with the metric given in Eq. (6.4) are needed to compute the Riemann tensor, which is used to compute the Ricci tensor and Ricci scalar. The Ricci tensor and Ricci scalar, in turn, fully determine the left-hand side of the Einstein's equation: $G_{\alpha\beta} \equiv R_{\alpha\beta} - \frac{1}{2}g_{\alpha\beta}\mathcal{R} = 0$.

It can be shown (best left for the exercises) that $G_{\alpha\beta} = 0$ leads to

$$-\frac{e^{N-P}}{r}\left(P' - \frac{1}{r}\right) - \frac{e^N}{r^2} = 0,$$
$$-\frac{N'}{r} - \frac{1}{r^2}\left(1 - e^P\right) = 0,$$
$$-\frac{1}{2}r^2 e^{-P}\left[N'' - \frac{1}{2}P'N' + \frac{1}{2}\left(N'\right)^2 + \frac{N' - P'}{r}\right] = 0. \tag{6.21}$$

These expressions yield

$$\frac{dP}{dr} = -\frac{dN}{dr} = \frac{1}{r}\left(1 - e^P\right), \tag{6.22}$$

which can be solved for P:

$$\int \frac{dP}{1 - e^P} = \int \frac{dr}{r}$$
$$\int \left(\frac{1 - e^P}{1 - e^P} + \frac{e^P}{1 - e^P}\right) dP = \ln Cr$$
$$P - \ln\left(1 - e^P\right) = \ln e^P - \ln\left(1 - e^P\right) = \ln\frac{e^P}{1 - e^P} = \ln Cr$$
$$\frac{e^P}{1 - e^P} = Cr \quad \rightarrow \quad e^P = \frac{Cr}{1 + Cr}, \tag{6.23}$$

where C is an integration constant. Solving for N we obtain

$$N = -P + const. \quad \rightarrow \quad e^N = e^{const.} e^{-P}, \tag{6.24}$$

which is valid in general (for all r). The requirement that the Minkowski metric must be recovered at large distances ($r \rightarrow \infty$):

$$\lim_{r\to\infty} g_{00} \rightarrow -1,$$
$$\lim_{r\to\infty} g_{11} \rightarrow 1, \tag{6.25}$$

sets the value of the integration constant: $\exp(const.) = 1$ or $const. = 0$. Therefore

$$N = -P$$
$$g_{00} = -e^N = -e^{-P} = -\frac{1}{g_{11}} = -\left(\frac{1 + Cr}{Cr}\right) = -\left(1 + \frac{1}{Cr}\right). \tag{6.26}$$

For weak gravitational fields, we derived in Eq. (4.98):

$$g_{00} = -\left(1 + \frac{2\Phi}{c^2}\right) = -\left(1 - \frac{2GM}{c^2 r}\right) = -\frac{1}{g_{11}} \quad \rightarrow \quad C = -\frac{c^2}{2GM}. \tag{6.27}$$

We arrive at the solution to the Schwarzschild problem, and the corresponding line element,

$$ds^2 = -\left(1 - \frac{2GM}{c^2 r}\right) c^2 dt^2 + \frac{dr^2}{1 - \frac{2GM}{c^2 r}} + r^2 d\theta^2 + r^2 \sin^2\theta \, d\phi^2, \tag{6.28}$$

in the Schwarzschild metric:

$$g_{\alpha\beta} = \begin{pmatrix} -\left(1 - \frac{2GM}{c^2 r}\right) & 0 & 0 & 0 \\ 0 & \left(1 - \frac{2GM}{c^2 r}\right)^{-1} & 0 & 0 \\ 0 & 0 & r^2 & 0 \\ 0 & 0 & 0 & r^2 \sin^2\theta \end{pmatrix}. \tag{6.29}$$

The derivation of the Schwarzschild metric has only two requirements for the distribution of matter in the problem – the matter:

- is spherically symmetric; and
- has zero density at the radius of interest.

Birkhoff's theorem: *Any* spherically symmetric *vacuum* ($T^{\alpha\beta} = 0$) solution of Einstein's field equations *must also be static* and agree with Schwarzschild's solution [14].

Therefore the spherically symmetric mass leads to the Schwarzschild metric *regardless* of whether the mass is static, collapsing, expanding, or pulsating. This, of course, refers to the field *outside* the mass, as first stated in the derivation, because we start with $T_{\alpha\beta} = 0$.

Two of the most important features of Newtonian gravity therefore apply to general relativity:

- The gravity of a spherical body appears to act from a central point mass;
- The gravitational field inside a spherical shell vanishes.

6.3. Schwarzschild radius, event horizon, and black holes

The Schwarzschild spacetime metric in Eq. (6.28) has a singularity when the denominator in the second term is equal to zero:

$$1 - \frac{2GM}{c^2 r} = 0. \tag{6.30}$$

This singularity of the Schwarzschild spacetime metric occurs at the *Schwarzschild radius*:

$$r_s = \frac{2GM}{c^2}. \tag{6.31}$$

This is also known as the *event horizon*, because events occurring inside it cannot propagate light signals to the outside.

Any body that is small enough to exist within its own event horizon is therefore disconnected from the rest of the Universe: its only physical manifestation is through its infinitely deep gravitational potential well,[3] which is what led to the adoption of the term *black hole* in the late 1960s.

[3] In the case of a rotating, charged isolated mass, the two additional properties that remain conserved are the angular momentum and the total charge.

For a body with mass equal to that of our Sun, the event horizon is equal to

$$r_s = \frac{2GM_\odot}{c^2} = \frac{2\left(6.67\times10^{-8}\right)\left(2\times10^{33}\right)}{\left(3\times10^{10}\right)^2} \approx 3\times10^5\ \text{cm} = 3\ \text{km}. \tag{6.32}$$

We can write the proper time in the Schwarzschild metric as

$$ds^2 = -d\tau^2 \quad \rightarrow \quad d\tau^2 = \left(1 - \frac{2GM}{c^2 r}\right)dt^2 - \frac{dr^2}{1-\frac{2GM}{c^2 r}} - r^2 d\theta^2 - r^2 \sin^2\theta\, d\phi^2, \tag{6.33}$$

where dt is the time interval according to an observer at $r \to \infty$, and $d\tau$ is the time interval measured by a *local* observer (in comoving coordinates, in which the Universe is static). As for the local observer the Universe is static, this means that $dr = 0$, hence,

$$dt^2 = \frac{d\tau^2}{1-\frac{2GM}{c^2 r}}. \tag{6.34}$$

This is *time dilation*: while the local observer near the black hole (at $r \gtrsim r_s$) sees nothing unusual about their time measurements ($d\tau$), the measurements of the observer at $r \to \infty$ would suggest that the local observer's clock runs slow by a factor $\left(1-\frac{2GM}{c^2 r}\right)^{-1/2}$. It becomes *infinitely* slow at the event horizon r_s. Therefore the inertial observer (at infinity) can *never* witness the infalling observer reach the event horizon.

6.4. Orbits in Schwarzschild's geometry

For the dynamics of black holes and the infalling matter that surrounds them (their accretion disks), it is important to quantify the motion of particles in their vicinity. We now present a brief exposition of the orbit theory near a black hole.

In order to compute orbits in Schwarzschild's geometry, we need to first compute the equations of motion. Due to spherical symmetry, the motion of the particle will be confined to a plane. This means that we can simplify the equations of motion by setting, without loss of generality, the polar angle coordinate to $\theta = \pi/2$ and $\dot{\theta} = 0$. The Lagrangian in Eq. (6.5) simplifies to

$$L = \frac{1}{2} g_{\alpha\beta}\dot{x}^\alpha \dot{x}^\beta = \frac{1}{2} g_{00} c^2 \dot{t}^2 + \frac{1}{2} g_{11}\dot{r}^2 + \frac{1}{2} g_{33}\dot{\phi}^2, \tag{6.35}$$

where $g_{00} = -\left(1 - 2GM/(c^2 r)\right)$, $g_{11} = \left(1 - 2GM/(c^2 r)\right)^{-1}$, and $g_{33} = r^2$.

Solving the Lagrange equations for coordinates ct and ϕ leads to, respectively

$$E = \dot{t} m_0 c^2 \left(1 - \frac{r_s}{r}\right),$$

$$L_\phi = \dot{\phi} m_0 r^2 \quad \rightarrow \quad l_\phi = \dot{\phi} r^2,$$

where L_ϕ is the angular momentum and $l_\phi \equiv L_\phi/m_0$ the angular momentum per unit mass and E is the energy per unit mass relative to infinity (*i.e.*, $\lim_{r\to\infty} E = 0$). Because it is always possible, even in generally curved spacetime, to choose a set of coordinates in which spacetime looks locally flat (as established in Section 4.5), the property of the four-momentum given for flat spacetime in Eq. (2.68) still applies:

$$\mathbf{p}\cdot\mathbf{p} = -km_0^2c^2, \tag{6.36}$$

where constant $k = 0$ for massless and $k = 1$ for massive particle with rest mass m_0. It can be shown (in the exercises) that the motion near the black hole can be described with

$$\left(\frac{dr}{d\tau}\right)^2 = \frac{E^2}{m_0^2c^2} - V_{\text{eff}}(r) \equiv \tilde{E} - V_{\text{eff}}(r), \tag{6.37}$$

where $V_{\text{eff}}(r)$ is a *relativistic effective potential* per unit mass

$$V_{\text{eff}}(r) \equiv c^2\left(k + \frac{l_\phi^2}{c^2r^2}\right)\left(1 - \frac{r_s}{r}\right), \tag{6.38}$$

after using the definition of the Schwarzschild radius for a mass M, given in Eq. (6.31); we also define the *effective energy* per unit mass $\tilde{E} \equiv E^2/(m_0^2c^2)$. Quantities $\tilde{E}$ and l_ϕ are constants, so-called *integrals of motion*, reflecting the conservation of energy and angular momentum. For each particle, the initial conditions $(r_0, \dot{r}_0, \phi_0, \dot{\phi}_0)$ set the integrals of motion $\tilde{E}$ and l_ϕ. The angular momentum l_ϕ dictates the shape of the relativistic effective potential. The interplay between the relativistic effective potential and the effective energy $\tilde{E}$ determines the type of orbit that the particle will have.

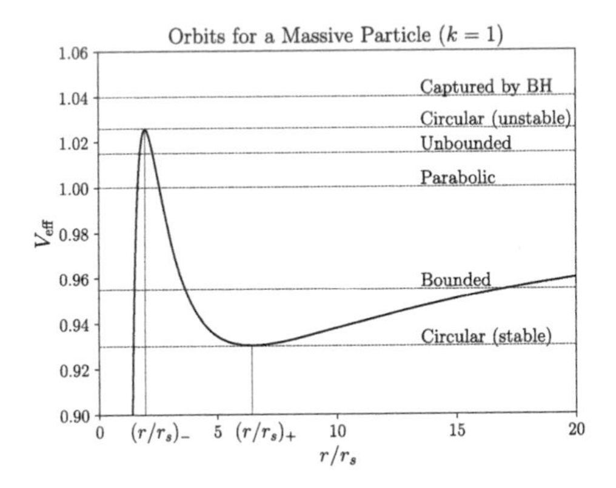

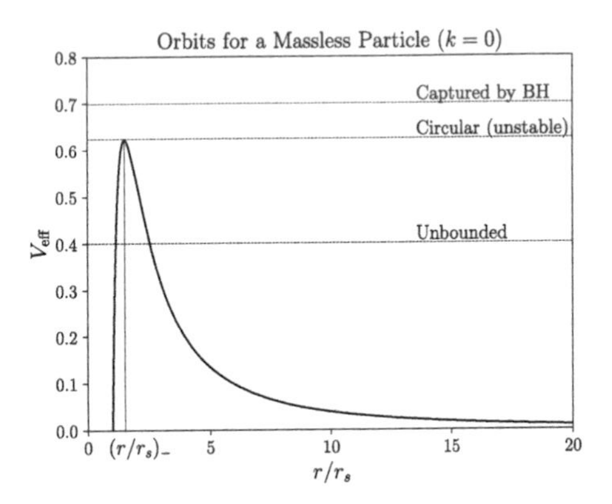

Figure 6.1 Orbits of massive (left panel) and massless (right panel) particles around a black hole. Here, $M = 1$, $G = 1$, $c = 1$, so $r_s = 2GM/c^2 = 2$, and $l_\phi = 4.1$. Dashed horizontal lines represent different orbital energy levels $\tilde{E}$.

The shape of the relativistic effective potential is given in Fig. 6.1. The two extrema of the potential are found by setting $dV_{\text{eff}}/dr = 0$:

$$\frac{dV_{\text{eff}}}{dr} = c^2\left(k\frac{r_s}{r^2} - \frac{2l_\phi^2}{c^2r^3} + \frac{3r_sl_\phi^2}{c^2r^4}\right) = 0 \quad \rightarrow \quad k\left(\frac{r}{r_s}\right)^2 - 2\left(\frac{l_\phi}{cr_s}\right)^2\left(\frac{r}{r_s}\right) + 3\left(\frac{l_\phi}{cr_s}\right)^2 = 0, \tag{6.39}$$

which after using the quadratic equation leads to

$$\left(\frac{r}{r_s}\right)_\pm = \left(\frac{l_\phi}{cr_s}\right)^2\left(1 \pm \sqrt{1 - \frac{3}{\left(\frac{l_\phi}{cr_s}\right)^2}}\right), \qquad \text{for massive particles } (k = 1), \tag{6.40}$$

$$\left(\frac{r}{r_s}\right)_- = \frac{3}{2}, \qquad \text{for massless particles } (k = 0). \tag{6.41}$$

These are the radii of the circular orbits. For massive particles, there are two circular orbits: an unstable circular orbit at $(r/r_s)_-$ and a stable circular orbit at $(r/r_s)_+$. There are *no* circular orbits if $l_\phi/(cr_s) < \sqrt{3}$. For massless particles, the circular orbit is unstable (called the *light ring*). The characteristics of orbits depend on the relationship between the particle's effective energy $\tilde{E}$ and the relativistic effective potential $V_{\text{eff}}(r)$:

- $\tilde{E} = V_{\text{eff,min}}$: stable, bound circular orbit.
- $V_{\text{eff,min}} < \tilde{E} < 1$: stable, bound elliptical orbit (bounded by two turning points at which $dr/d\tau = 0$ because $\tilde{E} = V_{\text{eff}}$).
- $\tilde{E} = 1$: unbound (parabolic) orbit barely escaping to infinity.
- $1 < \tilde{E} < V_{\text{eff,max}}$: unbound orbits.
- $\tilde{E} = V_{\text{eff,max}}$: unstable circular orbit.
- $\tilde{E} > V_{\text{eff,max}}$: particle coming from infinity will overcome the centrifugal (Coulomb) barrier and be captured by the black hole.

Exercises

1. Derive Eq. (6.21).
2. Derive Eq. (6.22).
3. Derive Eqs. (6.37) and (6.38).
4. **(Numerical)** Compute orbits of particles around an isolated massive object M in Schwarzschild geometry by numerically integrating Eqs. (6.37) and (6.38) for various initial conditions (r, ϕ) and values of the integrals of motion E and l_ϕ. Plot these orbits.

CHAPTER 7

The cosmological metric and Friedmann's equations

The most exciting phrase to hear in science, the one that heralds new discoveries, is not 'Eureka!' but 'That's funny...'

Isaac Asimov

The Big Picture: Earlier, we derived Einstein's field equations – a general-relativity analog to the Poisson equation – which describe how matter and radiation curve ambient spacetime. In this chapter, we are going to derive the Friedmann–Lemaître–Robertson–Walker metric (also known as the cosmological metric) for both flat and curved spacetimes in spherical coordinates, and look at the particular solutions for Universes with different contents.

7.1. Hubble's law and the evolving Universe

The "standard model" of the Universe is founded on the Cosmological Principle that states that our Universe is – at all times – *homogeneous* (the same from point to point) and *isotropic* (the same in all spatial directions) when viewed on the large scales (galaxies, galaxy clusters, galaxy super-clusters, etc., are considered "local inhomogeneities").

Consider four equally spaced observers along a line, as shown in Fig. 7.1. The velocity at which points d and a are moving from each other is then

$$v_{da} = 3v \propto R_{da} = 3R \quad \rightarrow \quad v_{da} = HR_{da}. \tag{7.1}$$

The assumption of isotropy of the standard model requires the parameter H to be independent of direction (angles of spherical coordinates):

$$H \neq H(\theta, \phi). \tag{7.2}$$

The assumption of homogeneity requires the parameter H to be independent of radius as well: $H \neq H(r)$.

We therefore arrive at **Hubble's Law** in vector form:

$$\boldsymbol{v} = H(t)\boldsymbol{r}. \tag{7.3}$$

Relativity and Cosmology
https://doi.org/10.1016/B978-0-44-323542-9.00015-8

The Hubble parameter $H(t)$ *is not a constant*, but is given in terms of the scale factor $a(t)$ as

$$H(t) \equiv \frac{\dot{a}(t)}{a(t)}. \tag{7.4}$$

Current measurements of the Hubble parameter are parameterized by the unitless parameter h:

$$H_0 = 100\ h\ \text{km s}^{-1}\ \text{Mpc}^{-1} = \frac{h}{9.78 \times 10^9\ \text{yr}} = 2.133 \times 10^{-33}\ h\ \text{eV}/\hbar, \tag{7.5}$$

with $h \approx 0.73$.[1]

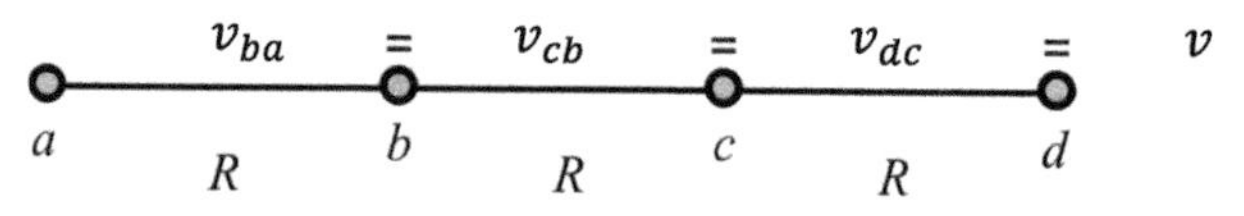

Figure 7.1 Cartoon depiction of Hubble's Law. Consider four points a, b, c, and d, located on the same line and equally spaced, as shown. If the entire line stretches uniformly, then each two neighboring points will move away from each other at the same speed v: $v_{ba} = v_{cb} = v_{dc} = v$. Then, the rate at which the two arbitrary points will move away from each other will be proportional to their distance: $v_{ca} = 2v$, $v_{da} = 3v \propto R_{da} = 3R$, hence, $v_{da} = HR_{da}$, or, simply, $v = HR$.

While the observations by Slipher [15] in 1912, and calculations by Friedmann [16] in 1922 and by Lemaître [17] in 1927, implying the expanding Universe predate Hubble's work [18] from 1929, the law remains largely known as the Hubble Law. In an encouraging recent attempt to make things right, the International Astronomial Union voted in 2018 to change its name to the *Hubble–Lemaître Law*.

The assumption of homogeneity of the standard model requires the Universe to have the same curvature everywhere (just like the 2D surface of a sphere has the same curvature everywhere).

Consider a 3-sphere[2] embedded in a 4D "hyperspace":

$$\left(x^0\right)^2 + \left(x^1\right)^2 + \left(x^2\right)^2 + \left(x^3\right)^2 = R^2, \tag{7.6}$$

[1] For the remainder of this textbook, we will give quantitative results in terms of this unitless parameter, and then explicitly state the value chosen. This is because two different approaches to measuring H_0 produce significantly different values, ranging from about 0.67 to about 0.73; see discussion on Hubble tension in Section 12.7.3.

[2] This is "just" one dimension up from a 2-sphere (like the surface of the Earth, for example) embedded in a 3D space.

where R is the radius of the 3-sphere. The distance between two points in 4D space is given by

$$dl^2 = \left(dx^0\right)^2 + \left(dx^1\right)^2 + \left(dx^2\right)^2 + \left(dx^3\right)^2. \tag{7.7}$$

Differentiating Eq. (7.6) and solving for dx^0, we obtain

$$dx^0 = -\frac{x^i dx^i}{\sqrt{R^2 - x^i x^i}}, \qquad \text{recall } i = 1, 2, 3 \tag{7.8}$$

so that Eq. (7.7) now reads

$$dl^2 = \frac{\left(x^i dx^i\right)^2}{R^2 - x^i x^i} + \left(dx^1\right)^2 + \left(dx^2\right)^2 + \left(dx^3\right)^2. \tag{7.9}$$

In spherical coordinates (r', θ, ϕ),

$$\begin{aligned} x^1 &= r' \sin\theta \cos\phi, \\ x^2 &= r' \sin\theta \sin\phi, \\ x^3 &= r' \cos\theta, \end{aligned}$$

hence,

$$\begin{aligned} dx^i dx^i &= dr'^2 + r'^2 d\theta^2 + r'^2 \sin^2\theta^2 d\phi^2, \\ x^i x^i &= r'^2, \\ x^i dx^i &= r' dr'. \end{aligned}$$

Now, we introduce a unitless *scale factor* $a(t)$ such that $R(t) \equiv a(t)R(t_0) \equiv a(t)R_0$, where $R_0 \equiv R(t_0)$ is the radius of the 3-sphere today $(t = t_0)$, so that presently $a(t_0) = a_0 = 1$. The line element on the 3-sphere embedded in a 4D space is then

$$\begin{aligned} dl^2 &= \frac{r'^2 dr'^2}{R^2 - r'^2} + dr'^2 + r'^2 d\theta^2 + r'^2 \sin^2\theta \, d\phi^2, \\ dl^2 &= \frac{R^2 dr'^2}{R^2 - r'^2} + r'^2 d\theta^2 + r'^2 \sin^2\theta \, d\phi^2, \\ dl^2 &= \frac{dr'^2}{1 - \left(\frac{r'}{aR_0}\right)^2} + r'^2 d\theta^2 + r'^2 \sin^2\theta \, d\phi^2. \end{aligned} \tag{7.10}$$

We could also have a negatively curved object (a "saddle") with $-a^2$ replacing a^2, or a flat (zero curvature, Euclidean) space with $a \to \infty$. We adopt a notation:

$$dl^2 = \frac{dr'^2}{1 - \frac{k}{R_0^2}\left(\frac{r'}{a}\right)^2} + r'^2 d\theta^2 + r'^2 \sin^2\theta \, d\phi^2, \tag{7.11}$$

$$ds^2 = -c^2 dt^2 + dl^2 = -c^2 dt^2 + \frac{dr'^2}{1 - \frac{k}{R_0^2}\left(\frac{r'}{a}\right)^2} + r'^2 d\theta^2 + r'^2 \sin^2\theta \, d\phi^2, \tag{7.12}$$

$$k \equiv \kappa R_0^2 = \begin{cases} 0 & \text{flat Universe (infinite, open),} \\ +1 & \text{positive-curvature Universe (finite, closed),} \\ -1 & \text{negative-curvature Universe (infinite, open),} \end{cases} \tag{7.13}$$

where κ is the *spacetime curvature* in units of length^{-2}, and k is unitless constant spacetime curvature. The two parameters, κ and k, have the same signs.

The flat, closed, and open manifolds are shown in Fig. 7.2. The most striking difference between the three geometries is illustrated by what happens to two lines that were initially parallel:

- Flat geometry $k = 0$: two parallel lines remain parallel forever and never cross;
- Elliptic geometry $k = +1$: two parallel lines move closer toward each other and eventually cross (like the lines of Earth's longitude being parallel at the equator, but crossing at the poles);
- Hyperbolic geometry $k = -1$: two parallel lines move farther away from each other.

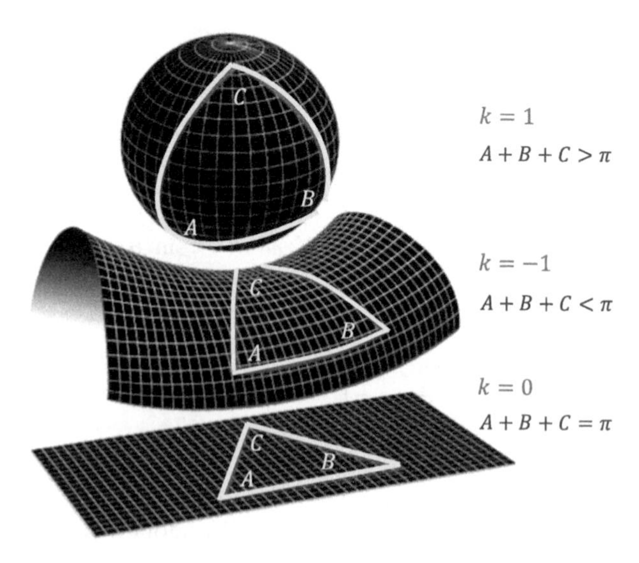

Figure 7.2 Closed ($k = 1$), open ($k = -1$), and flat geometries ($k = 0$).

If we define a new variable $r \equiv r'/a$, where, again, a is unitless and r has a unit of length, so that $dr \equiv dr'/a$, we finally obtain the spacetime interval

$$ds^2 = -c^2 dt^2 + a^2(t)\left(\frac{dr^2}{1-\kappa r^2} + r^2 d\theta^2 + r^2 \sin^2\theta\, d\phi^2\right) \tag{7.14}$$

for the Friedmann–Lemaître–Robertson–Walker metric in spherical coordinates:

$$g_{\alpha\beta} = \begin{pmatrix} -1 & 0 & 0 & 0 \\ 0 & \frac{a^2(t)}{1-\kappa r^2} & 0 & 0 \\ 0 & 0 & a^2(t) r^2 & 0 \\ 0 & 0 & 0 & a^2(t) r^2 \sin^2\theta \end{pmatrix}. \tag{7.15}$$

7.2. Friedmann's equations

We can now solve Einstein's field equations for the perfect fluid, as was first done by Alexander Friedmann [16] in 1922. All the calculations are done in a *comoving frame* where the spatial coordinates (r, θ, ϕ) remain fixed, while the entire coordinate system scales by $a(t)$:

$$u^0 = c = -u_0, \quad \text{and} \quad u^i = u_i = 0. \tag{7.16}$$

This means that the energy–momentum tensor is given by

$$T_{\alpha\beta} = \left(\rho + \frac{P}{c^2}\right) u_\alpha u_\beta + P g_{\alpha\beta}, \tag{7.17}$$

where ρ is mass density and P is pressure. Raising an index of Einstein's field equation (multiplying both sides by $g^{\alpha\nu}$),

$$R_{\nu\beta} - \frac{1}{2} g_{\nu\beta} \mathcal{R} = \frac{8\pi G}{c^4} T_{\nu\beta}, \tag{7.18}$$

we obtain

$$R^\alpha_\beta - \frac{1}{2}\delta^\alpha_\beta \mathcal{R} = \frac{8\pi G}{c^4} T^\alpha_\beta. \tag{7.19}$$

(Recall $g^{\alpha\nu} g_{\nu\beta} = \delta^\alpha_\beta$). After recalling the definition of the Ricci scalar from Eq. (5.27), $\mathcal{R} \equiv g^{\alpha\beta} R_{\alpha\beta} = R^\alpha_\alpha$, we can compute the Ricci scalar $\mathcal{R}$ either by finding a trace of the mixed Ricci tensor in Eq. (7.19), or by multiplying both sides of Eq. (7.18) by $g^{\nu\beta}$ to obtain

$$-\mathcal{R} = \frac{8\pi G}{c^4} T, \quad \text{where } T \equiv g^{\nu\beta} T_{\nu\beta} = T^\beta_\beta. \tag{7.20}$$

Combining this result with Eq. (7.19), we can rewrite Einstein's field equation as

$$R^{\alpha}_{\beta} = \frac{8\pi G}{c^4}\left(T^{\alpha}_{\beta} - \frac{1}{2}\delta^{\alpha}_{\beta}T\right). \tag{7.21}$$

To obtain a mixed tensor T^{α}_{β}, we raise an index of the covariant tensor $T_{\alpha\beta}$ given in Eq. (7.17),

$$T^{\nu}_{\beta} = g^{\nu\alpha}T_{\alpha\beta} = \left(\rho + \frac{P}{c^2}\right)u^{\nu}u_{\beta} + P\delta^{\nu}_{\beta}, \tag{7.22}$$

which we can then contract to obtain for the perfect fluid,

$$T = T^{\beta}_{\beta} = \left(\rho + \frac{P}{c^2}\right)(-c^2) + 4P = -\rho c^2 + 3P, \tag{7.23}$$

where we used $u^{\beta}u_{\beta} = -c^2$ and $\delta^{\beta}_{\beta} = 4$. This simplifies Eq. (7.21) as

$$R^{\alpha}_{\beta} = \frac{8\pi G}{c^4}\left[\left(\rho + \frac{P}{c^2}\right)u^{\alpha}u_{\beta} + \delta^{\alpha}_{\beta}P - \frac{1}{2}(-\rho c^2 + 3P)\delta^{\alpha}_{\beta}\right],$$
$$R^{\alpha}_{\beta} = \frac{8\pi G}{c^4}\left[\left(\rho + \frac{P}{c^2}\right)u^{\alpha}u_{\beta} + \frac{1}{2}(\rho c^2 - P)\delta^{\alpha}_{\beta}\right]. \tag{7.24}$$

Now, we need to compute the elements of the Ricci tensor on the left-hand side of the equation above by using its definition from the Riemann tensor in Eq. (5.22):

$$R_{\alpha\beta} = g^{\gamma\delta}R_{\gamma\alpha\delta\beta} = R^{\delta}_{\gamma\alpha\delta\beta}. \tag{7.25}$$

After a straightforward yet tedious calculation (perfect for the exercises), we obtain the components of the Ricci tensor,

$$\begin{aligned} R^{0}_{0} &= \frac{3\ddot{a}}{ac^2}, \\ R^{0}_{i} &= 0, \\ R^{i}_{j} &= \frac{1}{a^2}\left(\frac{a\ddot{a}}{c^2} + \frac{2\dot{a}^2}{c^2} + 2\kappa\right)\delta^{i}_{j}, \end{aligned} \tag{7.26}$$

where, again, $\dot{} \equiv \partial/\partial t$ and $\partial/\partial ct = (1/c)\partial/\partial t$.

The 00 component of Einstein's equation given in Eq. (7.24) becomes

$$\frac{3\ddot{a}}{ac^2} = \frac{8\pi G}{c^4}\left[-(\rho c^2 + P) + \frac{1}{2}(\rho c^2 - P)\right] \tag{7.27}$$

and finally takes on the form in which it is known as *Friedmann's second equation*:

$$\ddot{a} = -\frac{4\pi G}{3}\left(\rho + \frac{3P}{c^2}\right)a. \tag{7.28}$$

The ij component of Einstein's equation is (after recalling $u^i = u_i = 0$ as this is a comoving frame)

$$\frac{1}{a^2}\left(\frac{a\ddot{a}}{c^2} + \frac{2\dot{a}^2}{c^2} + 2\kappa\right)\delta^i_j = \frac{8\pi G}{c^4}\left[\frac{1}{2}(\rho c^2 - P)\delta^i_j\right],$$

$$\frac{1}{a^2}\left(\frac{a\ddot{a}}{c^2} + \frac{2\dot{a}^2}{c^2} + 2\kappa\right) = \frac{8\pi G}{c^4}\left[\frac{1}{2}(\rho c^2 - P)\right], \tag{7.29}$$

or

$$a\ddot{a} + 2\dot{a}^2 + 2\kappa c^2 = \frac{4\pi G}{c^2}(\rho c^2 - P)a^2. \tag{7.30}$$

It is, however, beneficial to further massage these basic equations into a set that is more easily solved. Solving Eq. (7.30) for $\ddot{a}$, we obtain

$$\ddot{a} = \frac{4\pi G}{c^2}(\rho c^2 - P)a - \frac{2\dot{a}^2}{a} + \frac{2\kappa c^2}{a}, \tag{7.31}$$

which can be combined with Eq. (7.28) to cancel out the P dependence to yield

$$\frac{16\pi G\rho a}{3} - \frac{2\dot{a}^2}{a} - \frac{2\kappa c^2}{a} = 0 \tag{7.32}$$

and finally takes on the form in which it is known as *Friedmann's first equation*:

$$\dot{a}^2 = \frac{8\pi G}{3}\rho a^2 - \kappa c^2. \tag{7.33}$$

Eqs. (7.28) and (7.33) are the basic equations connecting the scale factor a to ρ and P. To obtain a closed system of equations (uniquely solvable), we only need an *equation of state* $P = P(\rho)$, which relates P and ρ. Combined with the equation of state, the problem reduces to the system of three coupled equations combining three unknown functions for a, ρ, and P, which is mathematically well posed. In practice, as we will see next, the equation of state $P = P(\rho)$ for a particular 'state' of the Universe is directly inserted into two Friedmann's equations, thereby reducing to two equations for two unknowns: a and ρ.

When combined with Eq. (5.40) derived in the context of conservation of energy–momentum tensor, and the equation of state $P(\rho)$, we obtain a system of **Friedmann's equations**:

Friedmann's First Equation: $$\dot{a}^2 = \frac{8\pi G}{3}\rho a^2 - \kappa c^2, \tag{7.34a}$$

Friedmann's Second Equation: $$\ddot{a} = -\frac{4\pi G}{3}\left(\rho + \frac{3P}{c^2}\right)a, \tag{7.34b}$$

Friedmann's Third Equation: $$\dot{\rho} = -3\left(\rho + \frac{P}{c^2}\right)\frac{\dot{a}}{a}, \quad (7.34c)$$

Equation of State: $$P = P(\rho). \quad (7.34d)$$

Only the first two equations above are independent, and Friedmann's third equation can be derived from the first two for computational convenience (left for the exercises). Again, those two independent Friedmann's equations and one equation of state give three equations for the three unknowns – a, ρ, and P. This makes the system in Eq. (7.34) closed. We will solve this system of equations in the next chapter.

It is worthwhile to reiterate here the point we raised earlier. In Friedmann's equations above, ρ is the mass density. However, recalling the equivalence between energy and mass established by Einstein's revolutionary equation $E = mc^2$, the relationship between the mass density ρ and energy density u is easily recognized as $u = \rho c^2$. That means that Friedmann's equations can still be used for the evolution of energy density upon replacing $u = \rho c^2$. This is indeed what we are going to do later as we discuss the Universe when radiation energy density dominates. Inspecting dimensionality of the terms in the parentheses on the right-hand side of Friedmann's second and third equations provides an explanation for this substitution: energy density $u = \rho c^2$ has the same units as pressure P ($\mathrm{J/m^3 = N\ m/m^3 = N/m^2}$) (similarly, mass density ρ has the same units as P/c^2: ($\mathrm{kg/m^3 = kg\ m/m^4 = kg\ m\ s^2/(m^4\ s^2) = N\ s^2/m^4}$)).

7.3. A Newtonian analogy to Friedmann's equations

An intuitive and pedagogical understanding of Friedmann's equations can be gleaned from an approximate derivation based on a local Newtonian treatment.

The basic idea is to view the expanding Universe as the $r \to \infty$ limit of a spherically symmetric system that is undergoing an *homologous* expansion, so that, at any given instant, the density of the system is constant through space, although the density decreases over time.

Let us then consider the evolution of a fluid element located some distance a from an arbitrary origin $a = 0$ in the uniformly dense environment $\rho = const.$, shown in Fig. 7.3.

Newton's Second Law becomes

$$m\ddot{a} = F = -G\frac{mM(a)}{a^2} \quad \to \quad \ddot{a} = -G\frac{1}{a^2}\frac{4\pi}{3}a^3\rho \quad \to \quad \ddot{a} = -\frac{4G\pi}{3}\rho a, \quad (7.35)$$

where $\dot{} \equiv d/dt$ and ρ is the spatially uniform mass density (albeit decreasing with time) so that the mass enclosed within some radius is $M(a) = V(a)\rho = (4\pi a^3/3)\rho$. Here, we used Newton's Shell Theorem that states that a spherically symmetric object affects other objects gravitationally as if all of its mass were concentrated at its center. Eq. (7.35) is

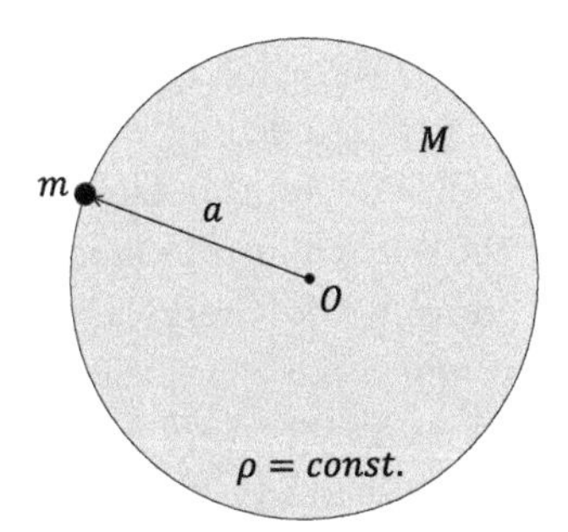

Figure 7.3 Newtonian derivation of Friedmann's equations. A temporal snapshot ($t = const.$) of a spherical mass M of uniform density $\rho(t = const.) = const.$ affecting another mass m some distance a away from its center O.

analogous to Friedmann's second equation,

$$\ddot{a} = -\frac{4G\pi}{3}\left(\rho + \frac{3P}{c^2}\right)a, \tag{7.36}$$

with one exception: the P term is missing, because we started with uniformly distributed matter without velocities creating internal pressure (dust approximation). Accounting for matter with random spatially isotropic velocities would lead to the P term analogous to that in Friedmann's second equation.

Energy conservation – the total energy, the sum of kinetic K and potential energies V is constant – requires

$$E = K + V = \frac{1}{2}m\dot{a}^2 - G\frac{mM(a)}{a} = const. \tag{7.37}$$

Canceling the test mass m and again using $M(a) = V(a)\rho = (4\pi a^3/3)\rho$, we obtain

$$\dot{a}^2 = \frac{8\pi G}{3}\rho a^2 + \frac{2E}{m}. \tag{7.38}$$

This equation is analogous to Friedmann's first equation. As before, Friedmann's third equation can be derived by combining the first two, and can be shown (in the exercises) to be another statement for energy conservation.

> Friedmann's first and third equations derive from energy conservation, and Friedmann's second equation derives from Newton's Second Law.

The equations derived here using Newtonian gravity are very similar to the proper relativistic equations. However, there *is* one important difference. In the context of general relativity, the constant k is related to the spatial curvature of the Universe. In the Newtonian analogy, there is no notion of curvature. Rather, the quantity $-\kappa/2$ is to be interpreted as the energy per unit mass for a particle at the point $a(t)$ in the expanding system: $-\kappa/2 = E/(mc^2)$. This interpretation has an important implication. If the energy is positive, so that the particle is unbound gravitationally, the Newtonian system will expand forever; if, instead, the energy is negative, so that the particle is bound gravitationally, the system will reach a point of maximum expansion and then begin to recontract. In other words, the sign of k has a crucial implication for the ultimate fate of the Universe. It will be seen later that, in the context of general relativity, this remains true: if the spatial curvature of the Universe is negative, the Universe will expand forever; however, if the spatial curvature is positive, the Universe will reach a point of maximum expansion and then recontract toward the Big Crunch.

Exercises

1. Starting with Friedmann's first and second equations, derive Friedmann's third equation.
2. Starting from the thermodynamic argument $dU = -PdV$, and the adiabatic condition $U = \rho c^2 V$, derive Friedmann's third equation. The solution will essentially restate the energy conservation derived for the Friedmann's first equation.
3. Find a geodesic equation for the expanding, flat Universe for which the spacetime line element is given in Eq. (7.14) by completing the following steps:
   - **a.** Identify coordinates (x^0, x^1, x^2, x^3).
   - **b.** Write down the spacetime metric $g_{\alpha\beta}$.
   - **c.** Write down the (alternative) Lagrangian that minimizes the line element and the Lagrange equations.
   - **d.** Solve the Lagrange equations for each coordinates x^ν to obtain a differential equation with $\ddot{x}^\nu$ on the left-hand side of the equation.
   - **e.** Identify all Christoffel symbols $\Gamma^\nu_{\gamma\delta}$ by comparing the differential equations in the previous part with the geodesic equation:

$$\ddot{x}^\nu = -\Gamma^\nu_{\gamma\delta}\dot{x}^\gamma\dot{x}^\delta.$$

CHAPTER 8

Solutions of Friedmann's equations

A man gazing at the stars is proverbially at the mercy of the puddles in the road.

Alexander Smith

The Big Picture: In the last chapter, we derived Friedmann's equations – a closed set of solutions of Einstein's equations that relate the scale factor $a(t)$, energy density ρ, and the pressure P for flat, open, and closed Universe (as denoted by the spacetime curvature constant $k=0, 1, -1$). In this chapter, we are going to solve Friedmann's equations for the matter-dominated and radiation-dominated Universe and obtain the form of the scale factor $a(t)$. We will also estimate the age of the flat Friedmann Universe. The glaring discrepancy between the predicted age of the flat, matter-dominated Universe and observations strongly implies that we are missing something essential. This, in turn, motivates our introduction of dark energy (the cosmological constant in Einstein's field equations), which is the topic of the next chapter.

8.1. Cosmological models

To generate a cosmological model, one must specify:

- An *equation of state*, relating ρ and P; and
- The value of the spacetime curvature constant κ, *i.e.*, which is positive for an open Universe, negative for a closed Universe, and zero for a flat Universe.

From the definition of the Hubble parameter H in Eq. (7.4),

$$H \equiv \frac{\dot{a}}{a} \quad \rightarrow \quad \dot{H} = -\frac{\dot{a}^2}{a^2} + \frac{\ddot{a}}{a} = -H^2 + \frac{\ddot{a}}{a} = -H^2\left(1 - \frac{\ddot{a}}{H^2 a}\right) \equiv -H^2\left(1+q\right), \tag{8.1}$$

we define a *deceleration parameter* q as

$$q \equiv -\frac{\ddot{a}}{H^2 a}. \tag{8.2}$$

8.1.1 Equation of state for the radiation-dominated Universe: perfect fluid

A radiation-dominated Universe is modeled by the perfect-fluid approximation with $P = u/3 = \rho c^2/3$ (for a derivation, see Appendix I).

Relativity and Cosmology
https://doi.org/10.1016/B978-0-44-323542-9.00016-X

Friedmann's third equation (Eq. (7.34c)) becomes

$$\dot{\rho}+3\left(\rho+\frac{1}{3}\rho\right)\frac{\dot{a}}{a}=\dot{\rho}+4\rho\frac{\dot{a}}{a}=0 \quad\rightarrow\quad a^4\dot{\rho}+4\rho\dot{a}a^3=0 \quad\rightarrow\quad \frac{d}{dt}\left(a^4\rho\right)=0$$
$$\rightarrow\quad a^4\rho=a_0^4\rho_0=const. \quad\rightarrow\quad \rho=\rho_0\left(\frac{a_0}{a}\right)^4 \quad\rightarrow\quad \rho=\rho_0 a^{-4}, \tag{8.3}$$

where we used $a_0=1$ in the last step.

The energy density of radiation evolves with the scale factor $a(t)$ as $u=u_0a^{-4}$.

8.1.2 Equation of state for the matter-dominated Universe: dust

A matter-dominated Universe is modeled by the (non-relativistic) dust approximation: $P=0$. Then, from Eq. (7.28), we have

$$\frac{\ddot{a}}{a}+\frac{4\pi G}{3}\rho=0, \tag{8.4}$$

and, in terms of H, from Eq. (8.2) $\ddot{a}/a=-H^2q$, hence,

$$-H^2q+\frac{4\pi G}{3}\rho=0. \tag{8.5}$$

Therefore

$$\rho=\frac{3H^2}{4\pi G}q. \tag{8.6}$$

Then, Friedmann's first equation, Eq. (7.34a), becomes

$$\left(\frac{\dot{a}}{a}\right)^2-\frac{8\pi G}{3}\rho=-\frac{\kappa c^2}{a^2},$$
$$H^2-2H^2q=-\frac{\kappa c^2}{a^2}, \tag{8.7}$$

hence,

$$-\kappa c^2=a^2H^2(1-2q). \tag{8.8}$$

Since both $a\neq 0$ and $H\neq 0$, for the flat Universe ($\kappa=0$), $q=1/2$ ($q>1/2$ for $\kappa>0$ and $q<1/2$ for $\kappa<0$). Combining the $q=1/2$ requirement for the flat Universe with Eq. (8.6) yields the *critical density* needed to produce the flat Universe:

$$\rho_{cr} = \frac{3H^2}{8\pi G}. \tag{8.9}$$

Currently, after using $h \approx 0.73$, it is (see Eq. (7.5))

$$\rho_{cr,0} = \frac{3H_0^2}{8\pi G} = \frac{3\left(\frac{h}{9.78\times10^9\ \mathrm{yr}}\right)^2\left(\frac{1\ \mathrm{yr}}{3600\times24\times365\ \mathrm{s}}\right)^2}{8\pi\left(6.67\times10^{-8}\mathrm{cm}^3\ \mathrm{g}^{-1}\ \mathrm{s}^{-2}\right)} = 1.87\times10^{-29}h^2\ \frac{\mathrm{g}}{\mathrm{cm}^3} \approx 10^{-29}\ \frac{\mathrm{g}}{\mathrm{cm}^3}.$$

That is, about five hydrogen atoms per cubic meter, or, perhaps more poetically, a small teardrop spread over the entire volume of the Earth.

It is important to note that the deceleration parameter q provides the relationship between the density of the Universe ρ and the critical density ρ_{cr} (after combining Eqs. (8.6) and (8.8)):

$$q = \frac{\rho}{2\rho_{cr}}. \tag{8.10}$$

Friedmann's third equation (Eq. (7.34c)) for the matter-dominated Universe ($P = 0$) becomes

$$\begin{aligned} &\dot{\rho} + 3\rho\frac{\dot{a}}{a} = 0 \quad \rightarrow \quad a^3\dot{\rho} + 3\rho\dot{a}a^2 = 0 \quad \rightarrow \quad \frac{d}{dt}\left(a^3\rho\right) = 0 \\ \rightarrow \quad &a^3\rho = a_0^3\rho_0 = const. \quad \rightarrow \quad \rho = \rho_0\left(\frac{a_0}{a}\right)^3 \quad \rightarrow \quad \rho = \rho_0 a^{-3}, \end{aligned} \tag{8.11}$$

where we used $a_0 = 1$ in the last step.

The energy density of matter evolves with the scale factor $a(t)$ as $\rho c^2 = \rho_0 c^2 a^{-3}$.

8.1.3 Curvature of the Universe

Rearranging Friedmann's first equation (Eq. (7.34a))

$$\frac{\dot{a}^2}{a^2} = \frac{8\pi G}{3}\rho - \frac{\kappa c^2}{a^2} \tag{8.12}$$

and recalling how radiation and matter contents depend on the scale factor a, $\rho_r = \rho_{r,0}(a_0/a)^4$ and $\rho_m = \rho_{m,0}(a_0/a)^3$, respectively (given in Eqs. (8.3) and (8.11)), we arrive at a more elucidating form of Friedmann's first equation for the Universe containing both radiation and matter:

$$\frac{\dot{a}^2}{a^2} = \frac{8\pi G}{3}\frac{\rho_{r,0}}{a^4} + \frac{8\pi G}{3}\frac{\rho_{m,0}}{a^3} - \frac{\kappa c^2}{a^2}, \tag{8.13}$$

where $\rho_{r,0}$ and $\rho_{m,0}$ are today's density content due to radiation and matter, respectively, and a_0 the scale of the Universe today – all constants. Now, we can clearly see the order in which portions of the right-hand side of the equation, describing the components affecting the evolution of the Universe, will dominate – for smaller values of a, the terms with the lower exponent will dominate first: when $a \ll 1$, $a^{-4} > a^{-3} > a^{-2}$. This means that as the Universe evolves, the most dominant component affecting its evolution will be, in chronological order: radiation, matter, and then the non-zero curvature (if present).

In the sections that follow, we will look into how the scale factor a evolves for various cosmologies, specified by the equation of state $P(\rho)$ and the curvature of spacetime $k = \kappa R_0^2 = \{-1, 0, 1\}$.

8.2. Flat Universe ($k = 0, q_0 = 1/2$)

8.2.1 Radiation-dominated Universe (perfect-fluid approximation)

For the radiation-dominated Universe, in Eq. (8.3) we derived that the energy density scales as $\rho = \rho_0 a^{-4}$. Then, Friedmann's first equation (Eq. (7.34a)) becomes

$$\frac{\dot{a}^2}{a^2} = \frac{8\pi G}{3}\rho = \frac{8\pi G}{3}\rho_0 a^{-4} \quad \rightarrow \quad \frac{da}{dt} = \sqrt{\frac{8\pi G\rho_0}{3}}\frac{1}{a}$$
$$\rightarrow \quad a da = \sqrt{\frac{8\pi G\rho_0}{3}}dt \quad \rightarrow \quad \int a da = \frac{1}{2}a^2 + K = \sqrt{\frac{8\pi G\rho_0}{3}}\, t. \tag{8.14}$$

At the Big Bang, $t = 0$, $a = 0$, so $K = 0$. As the Universe is flat, $\rho_0 = \rho_{cr}$. Therefore, using Eq. (8.9),

$$a = \left(\frac{32}{3}\pi G\rho_0\right)^{1/4} t^{1/2} = \left(\frac{32}{3}\pi G\rho_{cr}\right)^{1/4} t^{1/2} = \left(\frac{32}{3}\pi G\frac{3H_0^2}{8\pi G}\right)^{1/4} t^{1/2} = (2H_0)^{1/2} t^{1/2}.$$

The scale factor $a(t)$ for the flat, radiation-dominated Universe evolves as

$$a(t) = (2H_0)^{1/2} t^{1/2}. \tag{8.15}$$

Setting the scale factor a to the present-day value of $a_0 = 1$ allows us to find the age of the Universe t_0, which corresponds to the Hubble parameter H_0:

$$t_0 = \frac{1}{2H_0}. \tag{8.16}$$

Taking $H_0 = \frac{h}{9.78\times10^9\ \mathrm{yr}}$ and $h \approx 0.73$, we obtain

$$t_0 = \frac{9.78 \times 10^9\ \mathrm{yr}}{2h} \approx 6.7 \times 10^9\ \mathrm{yr} \equiv 6.7\ \mathcal{A}\ \ (\mathrm{aeon}). \tag{8.17}$$

The age of the flat, radiation-dominated Universe is about 6.7 billion years.

8.2.2 Matter-dominated Universe (dust approximation)

For the matter-dominated Universe, in Eq. (8.11) we derived that the mass density scales as $\rho = \rho_0 a^{-3}$. Then, Friedmann's first equation (Eq. (7.34a)) becomes

$$\begin{aligned} &\frac{\dot{a}^2}{a^2} = \frac{8\pi G}{3}\rho = \frac{8\pi G}{3}\rho_0 a^{-3} \quad \rightarrow \quad \frac{da}{dt} = \sqrt{\frac{8\pi G\rho_0}{3}}\frac{1}{a^{1/2}} \\ &\rightarrow \quad a^{1/2}da = \sqrt{\frac{8\pi G\rho_0}{3}}\,dt \quad \rightarrow \quad \int a^{1/2}da = \frac{2}{3}a^{3/2} + K = \sqrt{\frac{8\pi G\rho_0}{3}}\,t. \end{aligned} \tag{8.18}$$

As before, at the Big Bang, $t = 0$, $a = 0$, hence, $K = 0$. Upon adopting that the Universe is flat $\rho_0 = \rho_{\mathrm{cr}}$, we finally have

$$a = (6\pi G\rho_0)^{1/3}\, t^{2/3} = (6\pi G\rho_{\mathrm{cr}})^{1/3}\, t^{2/3} = \left(6\pi G\frac{3H_0^2}{8\pi G}\right)^{1/3} t^{2/3} = \left(\frac{3H_0}{2}\right)^{2/3} t^{2/3},$$

where, again, we use Eq. (8.9) in the second step.

The scale factor $a(t)$ for the flat, matter-dominated Universe evolves as

$$a(t) = \left(\frac{3H_0}{2}\right)^{2/3} t^{2/3}. \tag{8.19}$$

Setting the scale factor a to the present-day value of $a_0 = 1$, allows us to find the age of the Universe t_0, which corresponds to the Hubble parameter H_0:

$$t_0 = \frac{2}{3H_0}. \tag{8.20}$$

Taking $H_0 = \frac{h}{9.78\times10^9\ \mathrm{yr}}$ and $h \approx 0.73$, we obtain

$$t_0 = \frac{2 \times 9.78 \times 10^9\ \mathrm{yr}}{3h} \approx 8.9 \times 10^9\ \mathrm{yr} \equiv 8.9\ \mathcal{A}. \tag{8.21}$$

The age of the flat, matter-dominated Universe is about 8.9 billion years.

The ages of the flat, radiation-dominated and matter-dominated Universe are shown in Fig. 8.1.

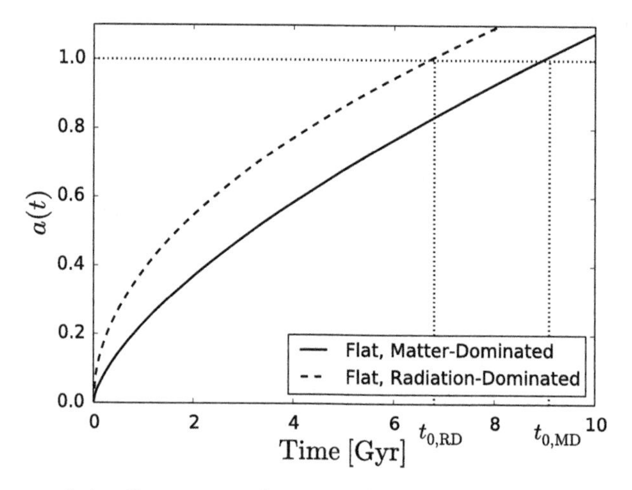

Figure 8.1 The age of the flat, matter-dominated ($t_{0,\mathrm{MD}}$) and radiation-dominated ($t_{0,\mathrm{RD}}$) Universe. $h = 0.73$ is used.

8.3. Closed Universe ($k = +1, q_0 > 1/2$)

8.3.1 Radiation-dominated Universe (perfect-fluid approximation)

For the radiation-dominated Universe, in Eq. (8.3) we derived that the energy density scales as $\rho = \rho_0 a^{-4}$. Then, Friedmann's first equation (Eq. (7.34a)) becomes

$$\frac{\dot{a}^2}{a^2} = \frac{8\pi G}{3}\rho - \frac{\kappa c^2}{a^2} = \frac{8\pi G}{3}\rho_0 a^{-4} - \frac{\kappa c^2}{a^2} \quad \rightarrow \quad \frac{da}{dt} = \sqrt{\kappa}c\sqrt{\frac{8\pi G\rho_0}{3a^2\kappa c^2} - 1}$$
$$\rightarrow \quad \int dt = \frac{1}{\sqrt{\kappa}c}\int \frac{da}{\sqrt{\frac{8\pi G\rho_0}{3a^2\kappa c^2} - 1}}.$$

Rewriting the integral above in terms of conformal time given by $d\eta \equiv dt/a$ (see Eq. (10.19)):

$$\int dt = \int a d\eta = \frac{1}{\sqrt{\kappa}c}\int \frac{da}{\sqrt{\frac{8\pi G\rho_0}{3a^2\kappa c^2} - 1}} \quad \rightarrow \quad \int d\eta = \frac{1}{\sqrt{\kappa}c}\int \frac{da}{\sqrt{\frac{8\pi G\rho_0}{3\kappa c^2} - a^2}}. \tag{8.22}$$

Using Eqs. (8.6)–(8.8), evaluated at $t = t_0$, we define the quantity that expresses the density ρ_0 in terms of the deceleration parameter at the present time q_0:

$$A_1 = \frac{8\pi G\rho_0}{3\kappa c^2} = \frac{2H_0^2 q_0}{\kappa c^2} = \frac{2H_0^2 q_0}{a_0^2 H_0^2 (2q_0 - 1)} = \frac{2q_0}{2q_0 - 1}, \tag{8.23}$$

after using $a_0 = 1$ in the last step. Then, the integral in Eq. (8.22) can be rewritten as

$$\eta - \eta_0 = \frac{1}{\sqrt{\kappa}c}\int_0^a \frac{d\tilde{a}}{\sqrt{A_1 - \tilde{a}^2}} = \frac{1}{\sqrt{\kappa}c}\sin^{-1}\left(\frac{a}{\sqrt{A_1}}\right). \tag{8.24}$$

The requirement[1] $\eta = 0$ at $a = 0$ sets $\eta_0 = 0$, hence, we have

$$a = \sqrt{A_1}\sin\left(\sqrt{\kappa}c\eta\right). \tag{8.25}$$

Now, $dt = a d\eta$, hence,

$$t - t_0 = \int a d\eta = \sqrt{A_1}\int \sin\left(c\sqrt{\kappa}\eta\right) d\eta = -\frac{\sqrt{A_1}}{\sqrt{\kappa}c}\cos\left(\sqrt{\kappa}c\eta\right). \tag{8.26}$$

The requirement $\eta = 0$ at $t = 0$ sets $t_0 = \sqrt{A_1}/c$.

We finally have for the closed, radiation-dominated Universe:

$$a = \sqrt{\frac{2q_0}{2q_0 - 1}}\sin\left(\sqrt{\kappa}c\eta\right),$$
$$\sqrt{\kappa}ct = \sqrt{\frac{2q_0}{2q_0 - 1}}\left(1 - \cos\left(\sqrt{\kappa}c\eta\right)\right). \tag{8.27}$$

Note that all individual terms in the equation above – a, $\sqrt{\kappa}ct$, $\sqrt{\kappa}c\eta$, and q_0 – are unitless.

8.3.2 Matter-dominated Universe (dust approximation)

For the matter-dominated Universe, in Eq. (8.11) we derived that the mass density scales as $\rho = \rho_0 a^{-3}$. Then, Friedmann's first equation (Eq. (7.34a)) becomes

$$\frac{\dot{a}^2}{a^2} = \frac{8\pi G}{3}\rho - \frac{\kappa c^2}{a^2} = \frac{8\pi G}{3}\rho_0 a^{-3} - \frac{\kappa c^2}{a^2} \quad \rightarrow \quad \frac{da}{dt} = \sqrt{\kappa}c\sqrt{\frac{8\pi G\rho_0}{3a\kappa c^2} - 1}$$

[1] More precisely, $\lim_{t\to 0}\eta = \int_0^t \frac{dt'}{a(t')} = 0$.

$$\rightarrow \quad \int dt = \frac{1}{\sqrt{\kappa}c}\int \frac{da}{\sqrt{\frac{8\pi G\rho_0}{3a\kappa c^2} - 1}}. \tag{8.28}$$

As before, we rewrite the integral above in terms of conformal time and define, after using Eqs. (8.6)–(8.8):

$$A \equiv \frac{4\pi G\rho_0}{3\kappa c^2} = \frac{H_0^2 q_0}{\kappa c^2} = \frac{H_0^2 q_0}{a_0^2 H_0^2(2q_0 - 1)} = \frac{q_0}{2q_0 - 1}. \tag{8.29}$$

Then,

$$\int d\eta = \frac{1}{\sqrt{\kappa}c}\int \frac{da}{\sqrt{\frac{8\pi G\rho_0}{3\kappa c^2}a - a^2}},$$

$$\rightarrow \quad \eta - \eta_0 = \frac{1}{\sqrt{\kappa}c}\int_0^a \frac{d\tilde{a}}{\sqrt{2A\tilde{a} - \tilde{a}^2}} = \frac{1}{\sqrt{\kappa}c}\left(\sin^{-1}\left(\frac{a-A}{A}\right) + \frac{1}{2}\pi\right). \tag{8.30}$$

However, the requirement $\eta = 0$ at $a = 0$ sets $\eta_0 = 0$, hence, we have

$$\frac{a-A}{A} = \sin\left(\sqrt{\kappa}c\eta - \pi/2\right) = -\cos\left(\sqrt{\kappa}c\eta\right) \quad \rightarrow \quad a = A\left(1 - \cos\left(\sqrt{\kappa}c\eta\right)\right). \tag{8.31}$$

As before, $dt = a d\eta$, hence,

$$t - t_0 = \int a d\eta = A\int \left(1 - \cos\left(\sqrt{\kappa}c\eta\right)\right) d\eta = A\left(\eta - \frac{1}{\sqrt{\kappa}c}\sin\left(c\sqrt{\kappa}\eta\right)\right). \tag{8.32}$$

However, the requirement $\eta = 0$ at $t = 0$ sets $t_0 = 0$.

We finally have for the closed, matter-dominated Universe:

$$a = \frac{q_0}{2q_0 - 1}\left(1 - \cos\left(\sqrt{\kappa}c\eta\right)\right),$$
$$\sqrt{\kappa}ct = \frac{q_0}{2q_0 - 1}\left(\sqrt{\kappa}c\eta - \sin\left(\sqrt{\kappa}c\eta\right)\right). \tag{8.33}$$

As we have kept the curvature term here (and in the derivation of the analogous equations for the open Universe below), the above expression captures the transition from the matter-dominated to the curvature-dominated Universe. In the case of non-flat spacetime geometry (open: $\kappa > 0$ or closed: $\kappa < 0$), the curvature term in Friedmann's first equation, Eq. (7.34a), scales as $\sim a^{-2}$, which will eventually always dominate the $\sim a^{-3}$ term due to the presence of matter (and, of course, the $\sim a^{-4}$ term due to radiation).

In both radiation- and matter-dominated closed Universes, the evolution is cycloidal – the scale factor grows at an ever-decreasing rate until it reaches a point at which the expansion is halted and reversed. The Universe then starts to compress and finally collapses in the Big Crunch.

8.4. Open Universe ($k = -1, q_0 < 1/2$)

8.4.1 Radiation-dominated Universe (perfect-fluid approximation)

For the radiation-dominated Universe, in Eq. (8.3) we derived that the energy density scales as $\rho = \rho_0 a^{-4}$. Then, Friedmann's first equation (Eq. (7.34a)) becomes

$$\frac{\dot{a}^2}{a^2} = \frac{8\pi G}{3}\rho + \frac{\kappa c^2}{a^2} = \frac{8\pi G}{3}\rho_0 a^{-4} + \frac{\kappa c^2}{a^2} \quad \rightarrow \quad \frac{da}{dt} = \sqrt{\kappa}c\sqrt{\frac{8\pi G\rho_0 a_0^4}{3a^2\kappa c^2} + 1}$$

$$\rightarrow \quad dt = \frac{1}{\sqrt{\kappa}c}\frac{da}{\sqrt{\frac{8\pi G\rho_0 a_0^3}{3a^2\kappa c^2} + 1}} \quad \rightarrow \quad \int dt = \frac{1}{\sqrt{\kappa}c}\int \frac{da}{\sqrt{\frac{8\pi G\rho_0 a_0^3}{3a^2\kappa c^2} + 1}}. \tag{8.34}$$

As before, we rewrite the integral above in terms of conformal time and quantity A_1 defined in Eq. (8.23):

$$\eta - \eta_0 = \frac{1}{\sqrt{\kappa}c}\int_0^a \frac{d\tilde{a}}{\sqrt{A_1 + \tilde{a}^2}} = \frac{1}{\sqrt{\kappa}c}\sinh^{-1}\left(\frac{a}{\sqrt{A_1}}\right). \tag{8.35}$$

Again, the requirement $\eta = 0$ at $a = 0$ sets $\eta_0 = 0$, so we have

$$a = \sqrt{A_1}\sinh\left(c\sqrt{\kappa}\eta\right). \tag{8.36}$$

As before, $dt = a d\eta$, hence,

$$t - t_0 = \int a d\eta = \frac{1}{\sqrt{\kappa}c}\sqrt{A_1}\cosh\left(\sqrt{\kappa}c\eta\right). \tag{8.37}$$

The requirement $\eta = 0$ at $t = 0$ sets $t_0 = \sqrt{A_1}/(\sqrt{\kappa}c)$.

We finally have for the open, radiation-dominated Universe:

$$a = \sqrt{\frac{2q_0}{1 - 2q_0}}\sinh\left(\sqrt{\kappa}c\eta\right), \tag{8.38}$$
$$\sqrt{\kappa}ct = \sqrt{\frac{2q_0}{1 - 2q_0}}\left(\cosh\left(\sqrt{\kappa}c\eta\right) - 1\right).$$

8.4.2 Matter-dominated Universe (dust approximation)

For the matter-dominated Universe, in Eq. (8.11) we derived that the mass density scales as $\rho = \rho_0 a^{-3}$. Then, Friedmann's first equation (Eq. (7.34a)) becomes

$$\frac{\dot{a}^2}{a^2} = \frac{8\pi G}{3}\rho + \frac{c^2}{a^2} = \frac{8\pi G}{3}\rho_0 a^{-3} + \frac{\kappa c^2}{a^2} \quad \rightarrow \quad \frac{da}{dt} = \sqrt{\kappa}c\sqrt{\frac{8\pi G\rho_0}{3ac^2}+1}$$
$$\rightarrow \quad dt = \frac{1}{\sqrt{\kappa}c}\frac{da}{\sqrt{\frac{8\pi G\rho_0}{3ac^2}+1}} \quad \rightarrow \quad \int dt = \frac{1}{\sqrt{\kappa}c}\int\frac{da}{\sqrt{\frac{8\pi G\rho_0}{3ac^2}+1}}. \tag{8.39}$$

Again, we rewrite the integral above in terms of conformal time:

$$\int d\eta = \frac{1}{\sqrt{\kappa}c}\int\frac{da}{\sqrt{\frac{8\pi G\rho_0}{3c^2}a + a^2}} \tag{8.40}$$

and use the quantity A defined in Eqs. (8.29) to obtain

$$\eta - \eta_0 = \frac{1}{\sqrt{\kappa}c}\int_0^a \frac{d\tilde{a}}{\sqrt{2A\tilde{a}+\tilde{a}^2}} = \frac{1}{\sqrt{\kappa}c}\cosh^{-1}\left(\frac{a+A}{A}\right). \tag{8.41}$$

However, the requirement $\eta = 0$ at $a = 0$ sets $\eta_0 = 0$, hence, we have

$$\frac{a+A}{A} = \cosh\left(\sqrt{\kappa}c\eta\right) \quad \rightarrow \quad a = A\left(\cosh\left(\sqrt{\kappa}c\eta\right) - 1\right). \tag{8.42}$$

Now, $dt = a d\eta$, hence,

$$t - t_0 = \frac{1}{\sqrt{\kappa}c}\int a d\eta = \frac{A}{\sqrt{\kappa}c}\int \left(\cosh\left(\sqrt{\kappa}c\eta\right) - 1\right)\, d\eta = \frac{A}{\sqrt{\kappa}c}\left(\sinh\left(\sqrt{\kappa}c\eta\right) - \sqrt{\kappa}c\eta\right). \tag{8.43}$$

However, the requirement $\eta = 0$ at $t = 0$ sets $t_0 = 0$.

We finally have for the open, matter-dominated Universe:

$$a = \frac{q_0}{1-2q_0}\left(\cosh\left(\sqrt{\kappa}c\eta\right) - 1\right), \tag{8.44}$$
$$\sqrt{\kappa}ct = \frac{q_0}{1-2q_0}\left(\sinh\left(\sqrt{\kappa}c\eta\right) - \sqrt{\kappa}c\eta\right).$$

In both radiation- and matter-dominated open Universes, the scale factor a increases without bound, which corresponds to the Universe that expands forever.

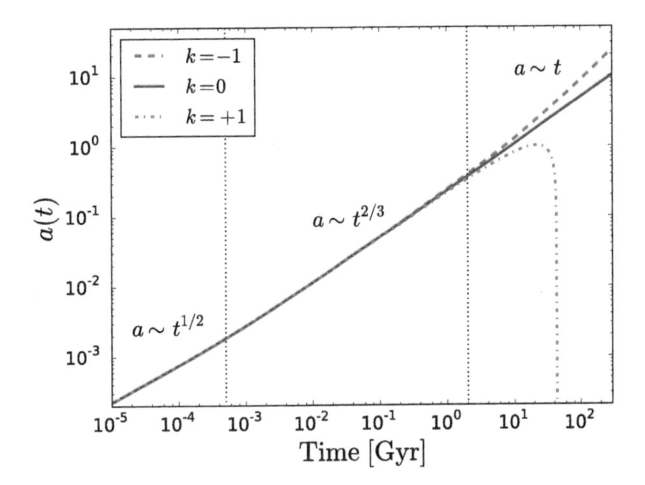

Figure 8.2 Evolution of the scale factor $a(t)$ for the flat ($k = 0$), closed ($k = +1$), and open ($k = -1$) Friedmann Universes. Radiation dominates early ($a \sim t^{1/2}$), followed by the matter-dominated epoch ($a \sim t^{2/3}$), and finally the curvature-dominated epoch ($a \sim t$). Vertical lines mark the transitions. $h = 0.73$ is used.

8.5. Big Bang: a radiation/matter-dominated singularity

At early times (for small values of η), the trigonometric and hyperbolic functions derived in the previous section to quantify the evolution of the scale factor a for different cosmologies, can be expanded in Taylor series (keeping only the first two terms):

$$\sin\left(\sqrt{\kappa}c\eta\right) \approx \sqrt{\kappa}c\eta - \frac{1}{6}\left(\sqrt{\kappa}c\eta\right)^3, \qquad \cos\left(\sqrt{\kappa}c\eta\right) \approx 1 - \frac{1}{2}\left(\sqrt{\kappa}c\eta\right)^2,$$

$$\sinh\left(\sqrt{\kappa}c\eta\right) \approx \sqrt{\kappa}c\eta + \frac{1}{6}\left(\sqrt{\kappa}c\eta\right)^3, \qquad \cosh\left(\sqrt{\kappa}c\eta\right) \approx 1 + \left(\sqrt{\kappa}c\eta\right)^2,$$

hence, to the leading term, the a and t dependence on η for the different curvatures is summarized in Tables 8.1 and 8.2.

Table 8.1 Scale factor $a(t)$ for the flat, closed, and open radiation-dominated Friedmann Universes. In the early Universe (small η), the evolution of the scale factor $a(t)$ is independent of the curvature k.

Curvature	**For all η**		**For small η**		
k	a	$\sqrt{\kappa}ct$	a	$\sqrt{\kappa}ct$	$a(t)$
0	$(32\pi G\rho_0/3)^{1/4}\,t^{1/2}$	–	$\propto t^{1/2}$	–	$\propto t^{1/2}$
+1	$\sqrt{\frac{2q_0}{2q_0-1}}\sin\left(\sqrt{\kappa}c\eta\right)$	$\sqrt{\frac{2q_0}{2q_0-1}}\left(1-\cos\left(\sqrt{\kappa}c\eta\right)\right)$	$\propto \eta$	$\propto \eta^2$	$\propto t^{1/2}$
−1	$\sqrt{\frac{2q_0}{1-2q_0}}\sinh\left(\sqrt{\kappa}c\eta\right)$	$\sqrt{\frac{2q_0}{1-2q_0}}\left(\cosh\left(\sqrt{\kappa}c\eta\right)-1\right)$	$\propto \eta$	$\propto \eta^2$	$\propto t^{1/2}$

Table 8.2 Scale factor $a(t)$ for the flat, closed, and open matter-dominated Friedmann Universes. In the early Universe (small η), the evolution of the scale factor $a(t)$ is independent of the curvature k.

Curvature	For all η		For small η		
k	a	$\sqrt{\kappa}ct$	a	$\sqrt{\kappa}ct$	$a(t)$
0	$(6\pi G\rho_0)^{1/3}\, t^{2/3}$	–	$\propto t^{2/3}$	–	$\propto t^{2/3}$
+1	$\frac{q_0}{2q_0-1}\left(1-\cos\left(\sqrt{\kappa}c\eta\right)\right)$	$\frac{q_0}{2q_0-1}\left(\sqrt{\kappa}c\eta-\sin\left(\sqrt{\kappa}c\eta\right)\right)$	$\propto \eta^2$	$\propto \eta^3$	$\propto t^{2/3}$
−1	$\frac{q_0}{1-2q_0}\left(\cosh\left(\sqrt{\kappa}c\eta\right)-1\right)$	$\frac{q_0}{1-2q_0}\left(\sinh\left(\sqrt{\kappa}c\eta\right)-\sqrt{\kappa}c\eta\right)$	$\propto \eta^2$	$\propto \eta^3$	$\propto t^{2/3}$

The moral of the story here is that singular behavior at early times is independent of the curvature of the Universe k. The Big Bang is a "radiation/matter-dominated singularity"!

The fact that the $k=\pm1$ solutions coincide with the $k=0$ solutions at early times, as shown in Tables 8.1 and 8.2 and illustrated in Fig. 8.2, is so important that it is worth laboring the point from another angle. Quite generally, Friedmann's first equation implies that

$$\dot{a}^2 = \frac{8\pi G}{3}\rho a^2 - \kappa c^2. \tag{8.45}$$

Suppose, however, that

$$\rho a^2 \propto a^{-n}, \tag{8.46}$$

where n is some constant ($n=1$ for the dust-filled Universe; $n=2$ for the radiation-filled Universe). It then follows that

$$\dot{a}^2 = \frac{C}{a^n} - \kappa c^2, \tag{8.47}$$

with C some constant. It is, however, apparent that, for $n>0$, a condition that holds for both radiation and dust, the right-hand side of the equation will always be dominated by the term C/a^n for $a\to 0$, so that the solution must always coincide with the $k=0$ solution. Similarly, it follows that, for sufficiently large values of a, the curvature term must dominate C/a^n, so that $\dot{a}^2=-\kappa c^2$, which explains the late-time behavior observed for a $k=-1$ Universe. The fact that the singular behavior at early times is essentially independent of the value of k is captured by the statement that "the Big Bang is a radiation/matter-dominated singularity."

Exercises

1. **(Numerical)** Use your favorite numerical tool to reproduce the curves in Fig. 8.2.

2. For the flat Friedmann's Universe, compute when the transition between the radiation-dominated and matter-dominated Universe happens.
3. Convert $\rho_{\mathrm{crit},0} = 3H_0^2/(8\pi G)$ into
   a. g cm^{-3};
   b. eV cm^{-3};
   c. protons cm^{-3};
   d. $M_\odot$ Mpc^{-3}.
   Use $H_0 = 100h$ km s^{-1} Mpc^{-1}, and $h = 0.73$. Give all answers in terms of the dimensionless parameter h first, and then use $h = 0.73$ for a numerical answer.

CHAPTER 9

Cosmological constant and the dark Universe

The simple is the seal of the true.

Subrahmanyan Chandrasekhar

The Big Picture: In the last chapter, we solved Friedmann's equations for the matter-dominated and radiation-dominated flat, open, and closed Universes and obtained the form of the scale factor $a(t)$. We computed the critical density needed to have a flat Universe at about 10^{-29} g cm^{-3}. We also estimated the age of the flat, matter-dominated Friedmann Universe to about 9 billion years. In this chapter, we are going to combine the information discovered by observations of the cosmic microwave background (CMB) radiation with the solutions of Friedmann's equations to present strong evidence for an additional vacuum energy and non-baryonic matter – dark energy and dark matter.

9.1. Age of a matter-dominated Friedmann Universe

At the present time, $t = t_0$ (age of the Universe), $a(t_0) = a_0$ and $q = q_0$, hence, Eq. (8.6) provides the link between the total current density of the Universe and the critical density:

$$q_0 = \frac{\rho_0}{2\rho_{cr}}. \tag{9.1}$$

Friedmann's equations provide the link between the age of the Universe t_0 and the present density of the Universe, given in terms of critical density ρ_{cr} via quantity q_0 (left for the exercises):

$$t_0 = \frac{1}{H_0}\begin{cases} \frac{1}{1-2q_0} - \frac{q_0}{(1-2q_0)^{3/2}}\cosh^{-1}\left(\frac{1-q_0}{q_0}\right) & \text{for } q_0 < \frac{1}{2}, \\ \frac{1}{1-2q_0} + \frac{q_0}{(2q_0-1)^{3/2}}\cos^{-1}\left(\frac{1-q_0}{q_0}\right) & \text{for } q_0 \geq \frac{1}{2}, \end{cases} \tag{9.2}$$

which, as we derived in Section 8.2.2, yields 8.9 billion years for the flat, matter-dominated Universe.

However, recent observations, such as Planck [19], suggest that the age of the Universe is

$$t_0 = 13.800 \pm 0.024\ \mathcal{A}, \tag{9.3}$$

https://doi.org/10.1016/B978-0-44-323542-9.00017-1

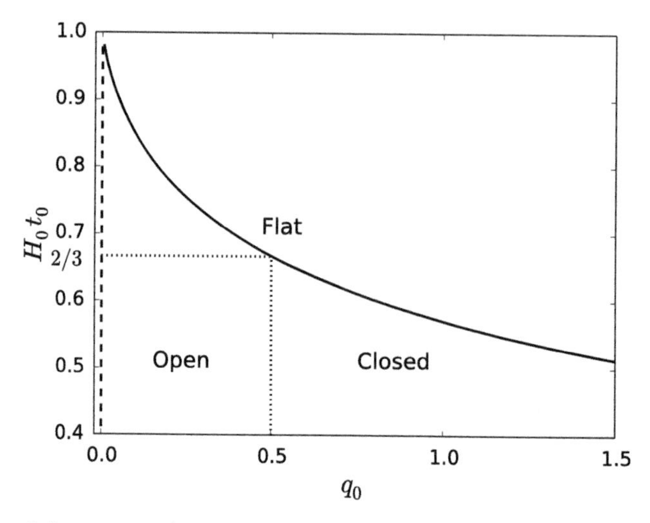

Figure 9.1 Age of the matter-dominated Friedmann Universe. Note that because $q_0 \propto \rho_0$, higher density implies a younger Universe. The dotted line corresponds to $t_0 = 8.9\ \mathcal{A}$, as computed in Section 8.2.2 for the flat Universe, and the dashed line to $t_0 = 13.8\ \mathcal{A}$, as observed.

which would – from Fig. 9.1 – imply that $q_0 \approx 0$, that is $\rho_0 \approx 0$ – there is *no* matter in the Universe! However, that is not the case – Planck data also indicate that the Universe is (very) nearly flat, so $q_0 = 1/2$. Hmmm... Something is wrong with the matter-dominated Friedmann Universe – it is missing most of its energy density.

9.2. Einstein's field equations revisited: cosmological constant

Einstein first introduced the cosmological constant Λ in his field equations in order to circumvent an embarrassing solution (at the time) – a non-steady-state Universe.

Einstein's equations with the cosmological constant are

$$R_{\alpha\beta} - \frac{1}{2} g_{\alpha\beta} \mathcal{R} + g_{\alpha\beta} \Lambda = \frac{8\pi G}{c^4} T_{\alpha\beta}, \tag{9.4}$$

or, alternatively,

$$R^{\alpha}_{\beta} - \frac{1}{2} \mathcal{R} + \Lambda = \frac{8\pi G}{c^4} T^{\alpha}_{\beta}, \tag{9.5}$$

$$R^{\alpha}_{\beta} - \frac{1}{2} \mathcal{R} = \frac{8\pi G}{c^4} \tilde{T}^{\alpha}_{\beta}, \tag{9.6}$$

where $\tilde{T}^{\alpha}_{\beta} = T^{\alpha}_{\beta} - \Lambda c^4/(8\pi G)$ and

$$\tilde{T}^{\alpha}_{\beta} = \begin{pmatrix} -\rho c^2 - \frac{\Lambda c^4}{8\pi G} & 0 & 0 & 0 \\ 0 & P - \frac{\Lambda c^4}{8\pi G} & 0 & 0 \\ 0 & 0 & P - \frac{\Lambda c^4}{8\pi G} & 0 \\ 0 & 0 & 0 & P - \frac{\Lambda c^4}{8\pi G} \end{pmatrix}. \tag{9.7}$$

The new energy–momentum tensor $\tilde{T}^{\alpha}_{\beta}$ reveals the nature of the cosmological constant Λ – it is a source of energy density and negative pressure (opposing the pressure of matter). Indeed, this is what led to the coining of the name *dark energy*.

The density of dark energy does not depend on the scale factor a. The conservation law (and also Friedmann's third equation) (Eq. 5.40),

$$\dot{\rho} + 3\left(\rho + \frac{P}{c^2}\right)\frac{\dot{a}}{a} = 0, \tag{9.8}$$

then implies that the equation of state for the dark energy is $P(\rho) \equiv -\Lambda c^4/(8\pi G) \equiv -\rho_\Lambda c^2 \equiv -u_\Lambda = const.$, where u_Λ is the dark-energy density and ρ_Λ its mass-density equivalent. More generally, since the equations of state for the matter is $P(\rho) = 0$ and radiation $P(\rho) = \rho_r c^2/3$, they can all be expressed as

$$P(\rho) = w\rho c^2, \tag{9.9}$$

where the parameter $w = -1$ for dark energy, $w = 0$ for matter and $w = 1/3$ for radiation.

Define the normalized (to critical mass density $\rho_{\mathrm{cr},0}$) energy densities due to radiation, matter, and dark-energy content, and their analog due to the curvature κ:

$$\begin{aligned}
\Omega_{\mathrm{r},0} &\equiv \frac{8\pi G}{3H_0^2}\rho_{\mathrm{r},0} = \frac{\rho_{\mathrm{r},0}}{\rho_{\mathrm{cr},0}} &&\rightarrow\quad H_0^2\Omega_{\mathrm{r},0} = \frac{8\pi G}{3}\rho_{\mathrm{r},0},\\
\Omega_{\mathrm{m},0} &\equiv \frac{8\pi G}{3H_0^2}\rho_{\mathrm{m},0} = \frac{\rho_{\mathrm{m},0}}{\rho_{\mathrm{cr},0}} &&\rightarrow\quad H_0^2\Omega_{\mathrm{m},0} = \frac{8\pi G}{3}\rho_{\mathrm{m},0},\\
\Omega_{\Lambda,0} &\equiv \frac{8\pi G}{3H_0^2}\rho_{\Lambda} = \frac{\rho_{\Lambda}}{\rho_{\mathrm{cr},0}} &&\rightarrow\quad H_0^2\Omega_{\Lambda,0} = \frac{8\pi G}{3}\rho_{\Lambda} = const.,\\
\Omega_{k,0} &\equiv -\frac{\kappa c^2}{H_0^2} &&\rightarrow\quad H_0^2\Omega_{k,0} = -\kappa c^2,
\end{aligned} \tag{9.10}$$

where we have used $\rho_{\Lambda,0} = \rho_\Lambda = const.$

The total density content of the Universe is

$$\rho = \rho_{\mathrm{r}} + \rho_{\mathrm{m}} + \rho_\Lambda = \rho_{\mathrm{r},0}a^{-4} + \rho_{\mathrm{m},0}a^{-3} + \rho_\Lambda, \tag{9.11}$$

or, in terms of normalized (by $\rho_{cr,0}$) densities,

$$\Omega_T \equiv \frac{\rho}{\rho_{cr,0}} = \Omega_{r,0}a^{-4} + \Omega_{m,0}a^{-3} + \Omega_{\Lambda,0}. \tag{9.12}$$

In terms of the quantity q, given in Eq. (8.10),

$$q_0 = \frac{\rho}{2\rho_{cr,0}} = \frac{\Omega_T}{2} \quad \rightarrow \quad \Omega_T = 2q_0. \tag{9.13}$$

Combining the definitions in Eq. (9.10), we can rewrite Friedmann's first equation (Eq. (7.34a)):

$$\left(\frac{\dot{a}}{a}\right)^2 - \frac{8\pi G}{3}\rho = -\frac{\kappa c^2}{a^2},$$
$$\left(\frac{\dot{a}}{a}\right)^2 - H_0^2\Omega_{r,0}a^{-4} - H_0^2\Omega_{m,0}a^{-3} - H_0^2\Omega_{\Lambda,0} = H_0^2\Omega_{k,0}a^{-2}. \tag{9.14}$$

After rearranging, we finally arrive at the Universe's budgetary content at the present moment (subscript 0):

$$\frac{1}{H_0^2}\left(\frac{\dot{a}}{a}\right)^2 - \Omega_{r,0}a^{-4} - \Omega_{m,0}a^{-3} - \Omega_{\Lambda,0} = \Omega_{k,0}a^{-2}. \tag{9.15}$$

Recalling the definition of the Hubble parameter $H \equiv \dot{a}/a$, it follows that, at the present time, $a = a_0 = 1$, the first term in the equation above is equal to unity, hence,

$$1 - \Omega_{r,0} - \Omega_{m,0} - \Omega_{\Lambda,0} = \Omega_{k,0}. \tag{9.16}$$

We can recast Eq. (8.8) that expresses Friedmann's first equation in terms of the quantity q:

$$-\kappa c^2 = a^2H^2(1 - 2q_0). \tag{9.17}$$

In terms of the total energy density budget,

$$-\kappa c^2 = a^2H^2(1 - \Omega_T), \tag{9.18}$$

which, once again, states that the critical density $\rho = \rho_{cr} \rightarrow \Omega_T = 1$, implies a flat Universe ($k = 0$, $\kappa = 0$).

9.3. Age of the Universe revisited

Now, consider a mixture of matter and dark energy only ($\rho_r = 0$):

$$\rho = \rho_m + \rho_\Lambda = \rho_{m,0} a^{-3} + \rho_{\Lambda,0}. \tag{9.19}$$

From Planck's observations the Universe appears flat, so $k = 0$ and $\Omega_{k,0} = 0$, which simplifies Eq. (9.16) to

$$\Omega_{m,0} + \Omega_{\Lambda,0} = 1 \quad \rightarrow \quad \Omega_{m,0} = 1 - \Omega_{\Lambda,0} \tag{9.20}$$

and Eq. (9.14) becomes

$$\left(\frac{\dot{a}}{a}\right)^2 = H_0^2 \left((1 - \Omega_{\Lambda,0}) \frac{1}{a^3} + \Omega_{\Lambda,0} \right). \tag{9.21}$$

Solving for $\dot{a}$, this becomes

$$\dot{a} = H_0 \sqrt{\frac{1 - \Omega_{\Lambda,0}}{a} + \Omega_{\Lambda,0} a^2} \tag{9.22}$$

and

$$\begin{aligned}
H_0 dt &= \frac{da}{\sqrt{\frac{1-\Omega_{\Lambda,0}}{a} + \Omega_{\Lambda,0} a^2}} \\
H_0 t_0 &= \int_0^1 \frac{da}{\sqrt{\frac{1-\Omega_{\Lambda,0}}{a} + \Omega_{\Lambda,0} a^2}} = \int_0^1 \frac{a^{1/2} da}{\sqrt{(1 - \Omega_{\Lambda,0}) + \Omega_{\Lambda,0} a^3}} \\
&= \frac{2}{3\sqrt{\Omega_{\Lambda,0}}} \ln \left(2 \left(\sqrt{\Omega_{\Lambda,0} a^3} + \sqrt{\Omega_{\Lambda,0}(a^3 - 1) + 1} \right) \right) \Bigg|_0^1 \\
&= \frac{2}{3\sqrt{\Omega_{\Lambda,0}}} \ln \left(\frac{1 + \sqrt{\Omega_{\Lambda,0}}}{\sqrt{1 - \Omega_{\Lambda,0}}} \right).
\end{aligned} \tag{9.23}$$

The age of the Universe with dark energy is

$$t_0 = \frac{2}{3 H_0 \sqrt{\Omega_{\Lambda,0}}} \ln \left(\frac{1 + \sqrt{\Omega_{\Lambda,0}}}{\sqrt{1 - \Omega_{\Lambda,0}}} \right). \tag{9.24}$$

As $\Omega_{\Lambda,0} \rightarrow 1$, $t_0 \rightarrow \infty$, some matter is needed to keep the age of the Universe finite. Hence, from the observations we obtained the age of the Universe, and from the w

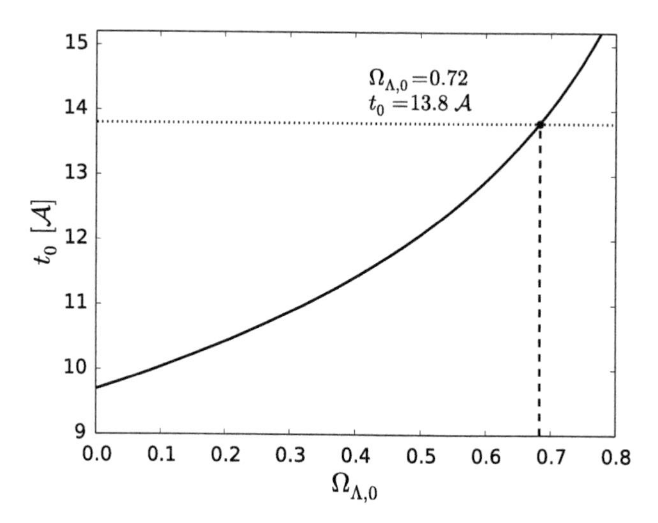

Figure 9.2 Age of the Universe with a cosmological constant Λ. The age of the Universe of 13.8 $\mathcal{A}$ corresponds to $\Omega_{\Lambda,0} \approx 0.6834$ and $h = 0.6727$. These are Planck collaboration parameters [19].

model for the equation of state of dark energy, we found, as shown in Fig. 9.2,

$$\Omega_{T,0} = \Omega_{m,0} + \Omega_{\Lambda,0} = 1, \qquad \text{the Universe is flat,} \tag{9.25}$$

$$\Omega_{\Lambda,0} = 0.72 \quad \rightarrow \quad \Omega_{m,0} = 0.28, \tag{9.26}$$

which means that

$$\frac{\Omega_{m,0}}{\Omega_T} \times 100\% = 28\% \quad \text{of the Universe is matter,}$$

$$\frac{\Omega_{\Lambda,0}}{\Omega_T} \times 100\% = 72\% \quad \text{of the Universe is dark energy.}$$

The Wilkinson Microwave Anisotropy Probe (WMAP) [20] and Planck [19] data also indicate that only 4% of the Universe is baryonic (normal) matter, and that 24% is in some other still unknown form (*dark matter*). The remainder of the energy budget of the Universe is provided by the *dark energy*. This means that we are completely ignorant of what 96% of the Universe is composed of!

9.4. Particle horizon revisited: the horizon problem

In Section 10.6, we will introduce the comoving horizon η. Its importance is in the fact that, under the standard cosmological model, the portions of the sky on our comoving horizon that are separated by more than η are *not causally connected* (there has not been an "exchange of information" between these regions ever in the history of the Universe). This means that, in the absence of interaction, different parts of the Universe should be in different states, leading to different initial conditions for the evolution of the Universe. The consequent evolution of different initial conditions would lead to different temperatures. However, they are all *very* similar, according to a remarkable isotropy of a few parts in 10^5 in the CMB radiation as measured by the WMAP and Planck probes! This is called the *horizon problem*.

9.5. Possible solutions to the horizon problem

It is widely believed that resolving the horizon problem requires allowing for all observable matter to have been causally connected early in the history of the Universe. There exist several competing theories that solve the horizon problem, each of them proposing various ways in which all observable matter was at some point in causal contact.

Cyclic cosmological models posit that the Universe goes through an infinite sequence of self-sustaining cycles: the *ekpyrotic* [21] model suggests that the Universe is caught in an eternal cycle of fiery birth, cooling, and rebirth, and that the Big Bang is really a big bounce, a transition from a previous epoch of contraction to the present epoch of expansion; *conformal cyclic cosmology* [22] proposes that the Universe continuously iterates: goes through infinite cycles with the future infinity of a previous iteration being identified with the Big Bang of the next.

For the remainder of this section, we focus on the *cosmological inflation*, or simply inflation, as the solution of the horizon problem.

9.5.1 Inflation

One way to solve the horizon problem is to allow all matter to interact, and therefore equilibrate to the same statistical properties, during the brief period of exponential expansion – *inflation* – immediately following the Big Bang. The effect of inflation is to make the current event horizon much larger than it would have been without inflation.

Consider an epoch during which some sort of inflation energy, perhaps dark energy, dominates other forms of energy density (radiation and matter): $\Omega_\Lambda \gg \Omega_r$, $\Omega_\Lambda \gg \Omega_m$, as well as the effects of spacetime curvature (recall Section 8.5: Big Bang is unaffected by curvature). Then, $\Omega_T \approx \Omega_\Lambda = 1$, and Eq. (9.21) becomes

$$\left(\frac{\dot{a}}{a}\right)^2 = H_0^2 \Omega_{\Lambda,0} = \frac{8\pi G}{3}\rho_\Lambda \equiv H^2 = const. \quad \rightarrow \quad \dot{a} = Ha \quad \rightarrow \quad a(t) \propto e^{Ht}, \tag{9.27}$$

where we used Eq. (9.10) to establish $H_0^2 \Omega_{\Lambda,0} = 8\pi G\rho_\Lambda/3 = const.$ The corresponding metric, for a so-called *de Sitter Universe*, is obtained by substituting $a(t) = \exp(Ht)$ in

Eq. (7.14):

$$ds^2 = -c^2 dt^2 + e^{2Ht}\left(dr^2 + r^2 d\theta^2 + r^2 \sin^2\theta \, d\phi^2\right). \tag{9.28}$$

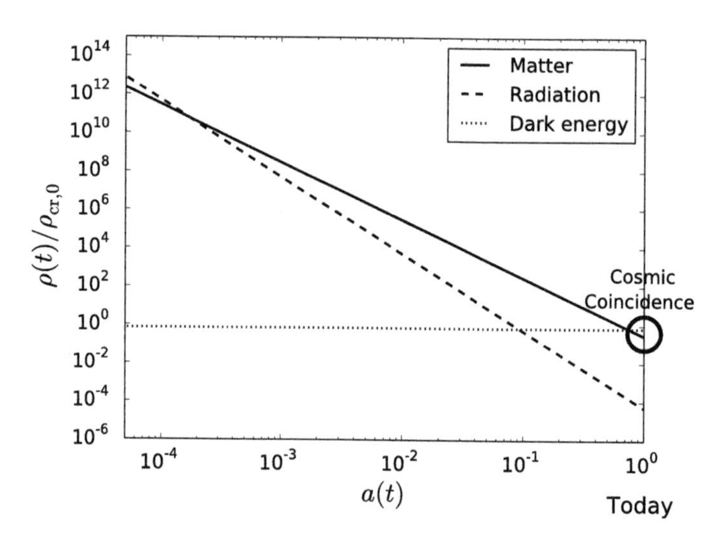

Figure 9.3 Relative importance of matter, radiation, and the cosmological constant Λ. The fact that today the cosmological constant and the matter content are of the same order of magnitude for the first time in the history of the Universe constitutes a so-called *cosmic coincidence problem.*

We are heading toward a de Sitter Universe, because the density of dark energy remains constant, while the matter density scales as a^{-3} and radiation density as a^{-4}, which makes the dark energy an ever-increasing part of the cosmic inventory (see Fig. 9.3).

The exponential expansion of the scale factor in Eq. (9.27) means that the physical distance between any two observers will eventually grow faster than the speed of light. At that point, those two observers will, of course, no longer be able to have any contact. Eventually, we will not be able to observe any galaxies other than the Milky Way and a handful of others in the gravitationally bound Local Group cluster of galaxies.

If we consider that the expansion occurred about the time when the grand unified theory (GUT) breaks down and the strong force becomes an independent force ("freezes out") at $t = t_{\mathrm{GUT}}$, then

$$H \approx \frac{1}{t_{\mathrm{GUT}}} \approx \frac{1}{10^{-36}\ \mathrm{s}} = 10^{36}\ \mathrm{s}^{-1}, \tag{9.29}$$

which is an *extremely fast* e-folding time, indicating a staggering rate of inflation. In just a few e-folding times, the Universe is already *huge*.

Inflation also solves the flatness problem

The WMAP and Planck data showed that the Universe is flat (or at least *very* nearly flat), *i.e.*, $\Omega_T \approx 1$. Why is this so? Why 1? Why not, say, 10^{-5} or 10^6? The standard model does not provide a reasonable explanation for the flat Universe. The problem is exacerbated, since the flat Universe is an *unstable fixed point*. This means that if the Universe started with $\Omega_T = 1$ *exactly*, it would remain so forever. If, however, the Universe was created with any other value of Ω_T, even one very close to but different from unity, the separation between the value of Ω_T and 1 would grow over time, presuming only that the scale factor a grows slower then linearly in time. Let us demonstrate this mathematically.

The first Friedmann equation (Eq. (7.34a)),

$$\dot{a}^2 = \frac{8\pi G}{3}\rho a^2 - \kappa c^2, \tag{9.30}$$

can be rewritten to yield

$$\rho = \frac{3}{8\pi G a^2}\left(\dot{a}^2 + \kappa c^2\right). \tag{9.31}$$

Dividing by the critical density,

$$\rho_{\mathrm{cr}} = \frac{3H^2}{8\pi G} \tag{9.32}$$

and recalling

$$\Omega_T = \frac{\rho}{\rho_{\mathrm{cr},0}} \tag{9.33}$$

leads to

$$\Omega_T = 1 + \frac{\kappa c^2}{H^2 a^2} \quad \rightarrow \quad \Omega_T - 1 = \frac{\kappa c^2}{\dot{a}^2}. \tag{9.34}$$

It is easily seen that if for $t \to 0$, $\dot{a} \to \infty$, then $\Omega_T - 1 \to 0$, and $\Omega_T \to 1$. The equation above implies that $\Omega_T \to 1$ very quickly, regardless of the value of the curvature parameter κ (recall, we noted earlier, in Section 8.5, that the curvature is relatively unimportant early in the history of the Universe – the behaviors of flat, closed, and open Universes are asymptotically identical at $t \to 0$). It also means that after inflation $\Omega_T = 1$ – the Universe is flat.

Let us assume $a = (t/t_0)^p$, and that $p < 1$, *i.e.*, a scales sublinearly. Then,

$$\dot{a} = t_0^{-p} p t^{p-1}, \tag{9.35}$$

so that

$$\Omega_{\rm T} - 1 = \frac{\kappa c^2}{\dot{a}^2} = \frac{\kappa c^2}{p^2} t_0^{2p} p^2 t^{2(1-p)} \equiv \tilde{\kappa} t^{2(1-p)}, \tag{9.36}$$

where $\tilde{\kappa} \equiv \kappa c^2 t_0^{2p}$. Finally, we obtain

$$\Omega_{\rm T} - 1 \propto t^{2(1-p)}, \tag{9.37}$$

so that

$$\begin{aligned} \Omega_{\rm T} - 1 &\to 0 \qquad \text{as } t \to 0 \\ \Omega_{\rm T} - 1 &\to \infty \qquad \text{as } t \to \infty. \end{aligned} \tag{9.38}$$

This means that the magnitude of $\Omega_{\rm T} - 1$ grows with increasing t. In other words, during the entire history of the Universe over which the scale factor a scales sublinearly, the Universe is growing increasingly non-flat (unless $\Omega_{\rm T}$ is exactly equal to unity). In the language of mathematics, $\Omega_{\rm T} = 1$ is an *unstable fixed point* for $p < 1$.

Eq. (9.37) holds a clue as to how to naturally obtain a flat Universe, in accordance with observations: change the dynamics so that $\Omega_{\rm T} = 1$ is a *stable fixed point*. All that is required is that the scale factor grows super-linearly (for examples $p > 1$ in the equations above or exponentially, as in the case of inflation). If one allows for the inflation energy so that a grows exponentially in time with $a(t) \propto \exp(Ht)$ (Eq. (9.27)), then

$$\dot{a} \propto He^{Ht}, \tag{9.39}$$

so that

$$\Omega_{\rm T} - 1 \propto \frac{\kappa c^2}{\dot{a}^2} \propto \frac{\kappa c^2}{H^2} e^{-2Ht}. \tag{9.40}$$

It follows that any initial deviation from unity is squashed exponentially. If, at some early time in its history, the Universe underwent a period of exponential expansion (inflation), any initial deviation from $\Omega_{\rm T} = 1$ would be reduced to the point *extremely* close to unity, so much so that even the prolonged subsequent evolution with $a \propto t^p$ with $p < 1$ would not drive it appreciably away from it. Therefore inflation solves the flatness problem.

Exercises

1. Derive Eq. (9.2) by using the results from Table 8.2 and Eq. (8.8).
2. Verify that for $q = 1/2$, Eq. (8.20) is recovered from Eq. (9.2).
3. For the flat Friedmann's Universe with a cosmological constant, compute when the transition between the matter-dominated and dark energy–dominated Universe happens.

4. Starting with Friedmann's first and second equations with the cosmological constant,

$$\left(\frac{\dot{a}}{a}\right)^2 = \frac{8\pi G}{3}\rho - \frac{\kappa c^2}{a^2} + \frac{\Lambda}{3},$$
$$\frac{\ddot{a}}{a} = -\frac{4\pi G}{3}\left(\rho + \frac{3P}{c^2}\right) + \frac{\Lambda}{3},$$

derive the corresponding Friedmann's third equation with the cosmological constant.

CHAPTER 10

Cosmic distances

Science never solves a problem without creating ten more.

George Bernard Shaw

The Big Picture: In this chapter, we are going to derive the redshift as a consequence of expansion of the Universe, and introduce the relevant distances associated with such an expanding Universe.

10.1. Redshift

If the wavelength of an emitted light is λ and if the observed wavelength is $\lambda_0 > \lambda$, then the line is said to be *redshifted* by a fraction z (the *redshift*) given by

$$z = \frac{\lambda_0 - \lambda}{\lambda}. \tag{10.1}$$

The cosmological redshift is a natural consequence of the expanding Universe: as the spatial extent of the Universe scales as a, the wavelength of a particle scales as

$$\frac{\lambda}{\lambda_0} = \frac{a}{a_0} = const. \quad \rightarrow \quad \lambda = \frac{\lambda_0 a}{a_0} \quad \rightarrow \quad \lambda = \lambda_0 a, \tag{10.2}$$

where we used the convention $a_0 = 1$. Combining this expression with Eq. (10.1) leads to the relationship between these two important cosmological quantities.

The redshift z and the scale factor a are related:

$$z = \frac{1-a}{a} = \frac{1}{a} - 1 \quad \rightarrow \quad a = \frac{1}{1+z}. \tag{10.3}$$

The gravitational redshift is observed when a receiver is located at a higher gravitational potential than the source. The physical explanation is that the particle loses a fraction of the energy (and hence increases its wavelength) by overcoming the difference in the potential (climbing out of the potential well).

https://doi.org/10.1016/B978-0-44-323542-9.00018-3

10.2. Proper distance

The distance from us to a galaxy at a coordinate r, as would be measured if the Universe were to instantaneously freeze and we used a measuring tape spanning the distance between us and the galaxy, is called the *proper distance*. The proper distance l_p is computed by setting $dt = 0$ (instantaneously frozen in time) and $d\theta = 0$, $d\phi = 0$ (straight radial distance) in Eq. (7.14) and integrating to obtain

$$l_p = \int_0^r \frac{a(t)}{\sqrt{1-\kappa r^2}}\, dr = \begin{cases} a(t)\, r & \kappa = 0 \text{ (infinite, open)}, \\ a(t)|\kappa|^{-1/2} \sin^{-1}\left(\sqrt{\kappa} r\right) & \kappa > 0 \text{ (finite, closed)}, \\ a(t)|\kappa|^{-1/2} \sinh^{-1}\left(\sqrt{\kappa} r\right) & \kappa < 0 \text{ (infinite, open)}. \end{cases} \tag{10.4}$$

Important note: for small r, $|\kappa|^{-1/2} \sin^{-1}\left(\sqrt{\kappa} r\right) \approx r$, $|\kappa|^{-1/2} \sinh^{-1}\left(\sqrt{\kappa} r\right) \approx r$. This means that local observations cannot distinguish between the three cases, $\kappa > 0$, $\kappa = 0$, or $\kappa < 0$, and therefore are inadequate for measuring the curvature of the Universe.

10.3. Comoving coordinates and comoving distance

General relativity states that the laws of physics are the same in any coordinates. However, some coordinates are easier to work with than others. One such set of coordinates are *comoving coordinates* in which an observer is comoving with the Hubble flow. Only for these observers in the comoving coordinates is the Universe isotropic (otherwise, portions of the Universe will exhibit a systematic bias: portions of the sky will appear systematically blue- or redshifted).

For the flat Friedmann–Lemaître–Robertson–Walker metric in Eq. (4.27) or its curved generalization introduced later in Eq. (7.15), the scale factor $a(t)$ multiplies its spatial part. If the scale factor increases with time, each observer sees other points in the Universe receding radially, according to Hubble's Law. A galaxy at comoving coordinates (r, θ, ϕ) remains at those coordinates, even as the spacetime around it expands according to the scale factor $a(t)$.

The comoving distance χ is the proper distance l_p at a given time divided by the scale factor $a(t)$. Therefore the comoving distance is $\chi = l_p/a(t)$, or

$$\chi = \begin{cases} r & \kappa = 0 \text{ (infinite, open)}, \\ |\kappa|^{-1/2} \sin^{-1}\left(\sqrt{\kappa} r\right) & \kappa > 0 \text{ (finite, closed)}, \\ |\kappa|^{-1/2} \sinh^{-1}\left(\sqrt{\kappa} r\right) & \kappa < 0 \text{ (infinite, open)}. \end{cases} \tag{10.5}$$

The comoving and proper distances are the same at the present time, when $a = a_0 = 1$, but will be different in the past and the future. The relationship between the proper and comoving distances is illustrated in Fig. 10.1.

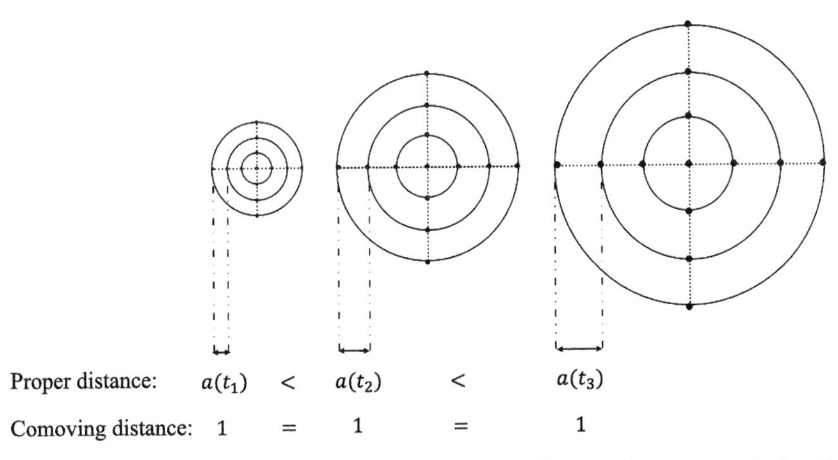

Figure 10.1 Comoving and proper (physical) distances. For an observer located at the center of the circle (stationary in the comoving coordinates), the Universe looks isotropic and homogeneous and it expands in all directions evenly. The comoving coordinates remain fixed, while the physical distance grows as $a(t)$. The two distances are related as $l_p = a\chi$, where l_p is the proper and χ is the comoving distance.

It is often useful to determine the comoving distance between a distant emitter ($t = t_e$, $a = a(t_e)$) and us ($t = t_0$, $a = a(t_0) = a_0 = 1$). In comoving coordinates, the distance to an object at a scale factor a (or alternatively redshift $z = 1/a - 1$) is

$$\chi \equiv c \int_{t_e}^{t_0} \frac{dt'}{a(t')} = c \int_{a}^{1} \frac{da'}{a'^2 H(a')}, \tag{10.6}$$

after the change of variables $da/dt = aH$, using the definition of the Hubble parameter $H \equiv \dot{a}/a$.

For the portion of the Universe that we can observe, which is to about $z \lesssim 20$, the radiation that dominated early on can be ignored. For the purely matter-dominated flat

Universe, we can combine the Hubble parameter $H = \dot{a}/a$ and Eq. (8.19),

$$a = \left(\frac{3H_0}{2}\right)^{2/3} t^{2/3}, \tag{10.7}$$

to obtain

$$H = \frac{\dot{a}}{a} = \frac{2\left(\frac{3H_0}{2}\right)^{2/3} t^{-1/3}}{3\left(\frac{3H_0}{2}\right)^{2/3} t^{2/3}} = \frac{2}{3t} = \frac{2}{3\frac{2}{3H_0}a^{3/2}} = H_0 a^{-3/2}. \tag{10.8}$$

This simplifies the integral in Eq. (10.6) to

$$\chi^{\mathrm{f,MD}}(a) = \frac{c}{H_0}\int_a^1 \frac{da'}{a'^{1/2}} = \frac{2c}{H_0} a'^{1/2}\Big|_a^1 = \frac{2c}{H_0}\left(1 - a^{1/2}\right), \tag{10.9}$$

where superscripts f and MD denote flat and matter-dominated Universe, respectively. In terms of the redshift z, Eq. (10.9) becomes (after recalling $z = 1/a - 1$)

$$\chi^{\mathrm{f,MD}}(z) = \frac{2c}{H_0}\left(1 - \frac{1}{\sqrt{1+z}}\right) \approx \frac{cz}{H_0}, \tag{10.10}$$

where we used the small redshift approximation $1/\sqrt{1+z} \approx 1 - z/2$ in the last step. This recovers Hubble's Law. For large redshift z, $1 \gg 1/\sqrt{1+z}$, so $\chi^{\mathrm{f,MD}}(z) \to 2/H_0$. The distance to the emitter is plotted in Fig. 10.2.

10.4. Angular diameter distance

Another important distance in astronomy is the *angular diameter distance*. It is determined by measuring the angle θ subtended by an object of known physical size l. Assuming that the angle is small, it is given by

$$d_A = \frac{l}{\theta}. \tag{10.11}$$

To compute the angular diameter distance in an expanding Universe, we express the quantities l and θ in comoving coordinates. The comoving size of a distant[1] object of physical size l is simply $l_c = l/a$. The angle subtended in the flat Friedmann Universe is the ratio of the object's comoving size and its comoving distance to the object, given in Eq. (10.6):

$$\theta = \frac{l_c}{\chi(a)} = \frac{\frac{l}{a}}{\chi(a)} = \frac{l}{a\chi(a)}. \tag{10.12}$$

[1] The object has to be far enough away from the observer so as to not be gravitationally bound to it. Gravitationally bound systems – from stellar binaries up to the superclusters – are "pinched off" from the inflation of the Universe.

Combining the previous two equations, and using Eq. (10.3), we finally have

$$d_A = \frac{l}{\frac{l}{a\chi(a)}} = a\chi = \frac{\chi}{1+z}.$$

The *angular diameter distance* d_A is

$$d_A = a\chi = \frac{\chi}{1+z}, \tag{10.13}$$

where χ is the comoving distance.

For small redshift z, $d_A^{\mathrm{f,MD}} \approx \chi$. At large z, $d_A^{\mathrm{f,MD}} \approx \chi/z$, the angular diameter distance decreases with redshift z. This means that in the flat Universe, objects at large redshifts appear larger than they would at intermediate redshifts! This is illustrated in Fig. 10.2.

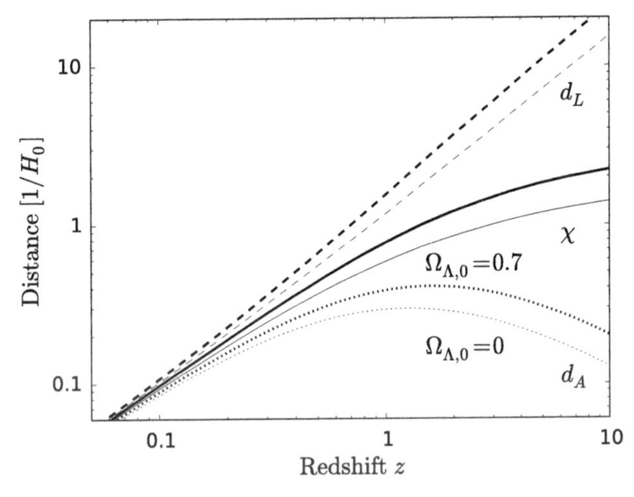

Figure 10.2 Three distances measures in a flat expanding matter-dominated Universe with no dark energy (thin lines) and a Universe with matter and dark energy corresponding to $\Omega_{\Lambda,0} = 0.7$ (thick lines). Solid lines correspond to the comoving distance χ, dotted lines to angular diameter distance d_A, and dashed lines to luminosity distance d_L.

10.5. Luminosity distance

In astronomy, distances can be inferred by measuring the flux from an object of known luminosity ("standard candles"). Flux and luminosity are related through

$$f \equiv \frac{L}{4\pi d^2}, \tag{10.14}$$

since the total luminosity through a spherical constant with area $4\pi d^2$ is constant. The total luminosity is defined as the amount of energy radiated per unit time (therefore, luminosity is power – an intrinsic property). This means $L \equiv dE/dt$. Assuming that, without loss of generality, all the N photons radiated have the same frequency ν (wavelength λ), then the luminosity becomes $L = (\hbar/\lambda)(dN/dt)$. In comoving coordinates $\lambda_c = \lambda/a$ and the t-derivative is replaced by the η-derivative (recall $dt = ad\eta$), hence,

$$L(\chi) = \frac{\hbar}{\lambda_c}\frac{dN}{d\eta} = a\frac{\hbar}{\lambda}\frac{dN}{dt}a = \frac{\hbar}{\lambda}\frac{dN}{dt}a^2 = La^2. \tag{10.15}$$

Then, the observed flux is

$$f = \frac{La^2}{4\pi\chi^2} = \frac{L}{4\pi\left(\frac{\chi}{a}\right)^2} \equiv \frac{L}{4\pi d_L^2}, \tag{10.16}$$

where

$$d_L \equiv \frac{\chi}{a}, \tag{10.17}$$

is the *luminosity distance*.

All three distances discussed in this chapter – comoving, angular diameter, and luminosity – are larger in a Universe with a cosmological constant than in one without.

Important note: Reliable measurements of these distances, when combined with accurate measurements of the redshift z, can provide a constraint on the energy density of the dark energy Ω_Λ (as will be discussed later in more detail).

10.6. Particle horizon

The particle horizon (also known as *the cosmological horizon* or *the comoving horizon*) is defined as the total portion of the Universe visible to the observer. It represents a sphere with the radius equal to the distance the light could have traveled (in the absence of interactions) since the Big Bang ($t = 0$). (See Fig. 10.3.)

In time dt, light travels a comoving distance:

$$cd\eta \equiv \frac{dx}{a} = \frac{cdt}{a}, \tag{10.18}$$

hence,

$$\eta \equiv \int_0^t \frac{dt'}{a(t')}. \tag{10.19}$$

η is called the *conformal time* that has passed since the Big Bang to some time t.

The particle horizon is a sphere of radius equal to the conformal time times the speed of light:

$$d_p(t) = c\eta = c\int_0^t \frac{dt'}{a(t')}. \tag{10.20}$$

The particle horizon is important because no information could have traveled further than $d_p(t)$ since the Big Bang.

For a spatially flat, matter-dominated Universe, as was discussed in detail in Chapter 8, the scale factor behaves as in Eq. (8.19):

$$a \propto t^{2/3}. \tag{10.21}$$

At present ($t = t_0$), the scale factor is $a(t_0) \equiv a_0 = 1$, so Eq. (10.21) can be expressed as

$$a(t) = \left(\frac{t}{t_0}\right)^{2/3}. \tag{10.22}$$

Then, the conformal time for a flat, matter-dominated Universe can be computed to be

$$\eta_0 \equiv \eta(t_0) = \int_0^{t_0} \frac{dt'}{\left(\frac{t'}{t_0}\right)^{2/3}} = t_0^{2/3}(3t_0^{1/3}) = 3t_0 \approx 41.1 \text{ billion years}, \tag{10.23}$$

so that the radius of the sphere defining the particle horizon today is[a]

$$d_p(t_0) = c\eta_0 \approx 41.1 \text{ billion light years}. \tag{10.24}$$

[a] For the radius of the particle horizon of a flat Universe with both matter and a cosmological constant, see the exercises below.

As conformal time η is a monotonically increasing variable of time t, it can be used as an independent variable when discussing the evolution of the Universe (just like the time t, temperature T, redshift z, and the scale factor a). In some approximations, Eq. (10.19)

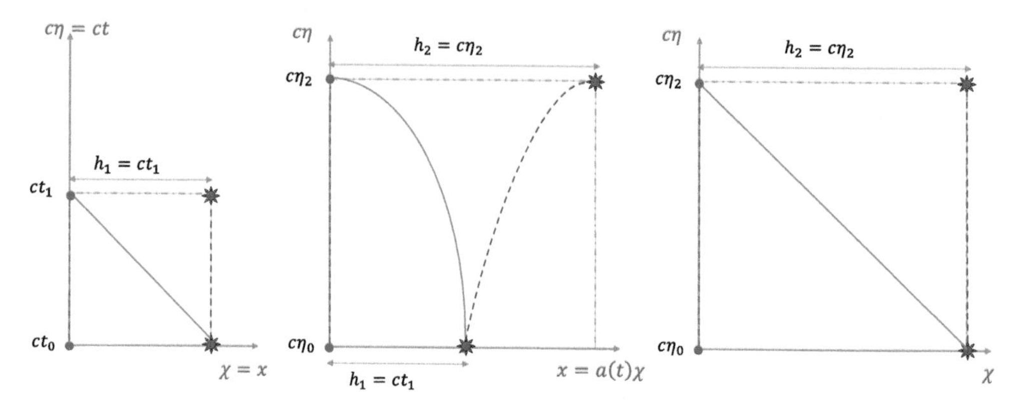

Figure 10.3 The spacetime diagrams for the particle horizon for a steady-state Universe (left panel) and expanding Universe (middle and right panels). At time $t = t_0$, a star is at the edge of the particle horizon for an observer located at the dot (both at rest in comoving coordinates). Their respective worldlines are shown in dashed lines. The star emits light, which follows the solid curve. For a steady-state Universe ($a(t) = 1$), we recover the traditional, special-relativistic Minkowski spacetime diagrams in which the trajectory of light is a straight line (left panel). In an expanding Universe, light has to travel farther to reach the observer, resulting in a larger particle horizon. For an expanding Universe, the spatial coordinate $a(t)x$ changes with time (it becomes longer as we march forward). In the spacetime diagrams of physical distance x and conformal time η (middle panel), the shape of the star's worldline is now curved, as is that of the light it emits. In the spacetime diagrams of comoving distance ξ and conformal time η (right panel), the shape of the star's worldline is again a straight line, as is that of the light it emits.

above can be analytically solved. For instance, in a matter-dominated Universe, $\eta \propto a^{1/2}$, and in a radiation-dominated Universe, $\eta \propto a$.

Exercises

1. Consider the size of the particle horizon estimated for a flat, matter-dominated Universe in Eq. (10.23). Compute by how much the estimate would change if we accounted for radiation-dominated early period from $t = 0$ to $t \approx t_0/200\,000$.
2. Compute by how much the estimate in the previous exercise would change if we accounted for the dark energy-dominated late period.
3. **(Numerical)** Use your favorite plotting tool to reproduce the curves in Fig. 10.2.

4. Assume the Universe today is flat with both matter ($\Omega_{m,0}$) and a cosmological constant (Ω_Λ).
   a. **(Numerical)** Compute the conformal age or the particle horizon of the Universe and plot your result as a function of $\Omega_{m,0}$. Numerical integration may be necessary. Plot it using your favorite electronic graphing tool.
   b. What is the current particle horizon size for a Universe with $\Omega_{m,0} = 0.28$? Use $H_0 = 100h$ km s^{-1} Mpc^{-1}, where $h \approx 0.73$.
   c. How would the previous estimate change if the value of $h \approx 0.67$ were used?
   d. What is the mass contained within the current horizon in solar masses for $\Omega_{m,0} = 0.28$?

CHAPTER 11

Summary of the foundations of cosmology

Shall I refuse my dinner because I do not fully understand the process of digestion?

Oliver Heaviside

The Big Picture: In the past several chapters, we introduced and developed the basic ideas of general relativity as they pertain to the understanding of the Universe on the largest scales. We derived the equations of general relativity that describe the kinematics in a curved spacetime – geodesic equation and the Einstein's field equations. Solving Einstein's field equations, both with and without the cosmological constant, leads to different cosmologies, which depend on both the curvature – flat, closed, and open – and the content of the Universe – matter, radiation, and dark energy. Here, we review these concepts.

11.1. General relativity: kinematics in curved spacetime

General relativity describes the kinematics in curved spacetime through two equations:

- *Geodesic equation*: How a particle moves in curved spacetime (general relativistic analogy to Newton's First and Second Laws in flat Euclidean space)

$$\ddot{x}^{\nu} = -\Gamma^{\nu}_{\gamma\delta}\dot{x}^{\gamma}\dot{x}^{\delta}. \tag{11.1}$$

- *Einstein's field equations*: How mass and energy distort (curve) spacetime (general relativistic analogy to Poisson's equation that describes how mass distribution creates a force field in Newtonian mechanics)

$$R^{\alpha}_{\beta} - \frac{1}{2}\mathcal{R} + \Lambda = \frac{8\pi G}{c^4}T^{\alpha}_{\beta}, \tag{11.2}$$

where Λ is a cosmological constant corresponding to dark energy.

Relativity and Cosmology
https://doi.org/10.1016/B978-0-44-323542-9.00019-5

To solve Einstein's field equations, we use the Frieadmann–Lemaître–Robertson–Walker metric,

$$ds^2 = -c^2 dt^2 + a^2(t)\left(\frac{dr^2}{1-\kappa r^2} + r^2 d\theta^2 + r^2 \sin^2\theta \, d\phi^2\right), \tag{11.3}$$

with (possibly) evolving space (through the scale factor $a(t)$, which does not have to be time dependent *a priori*). This metric was derived by invoking the *Cosmological Principle*, which assumes two important approximations – that the Universe is: (1) isotropic (the same in all directions) and (2) homogeneous (the same everywhere).

These approximations make the metric tensor diagonal and significantly reduce the complexity of Einstein's field equations, thereby leading to a simplified set of Friedmann's equations:

Friedmann's First Equation: $$\dot{a}^2 = \frac{8\pi G}{3}\rho a^2 - \kappa c^2, \tag{11.4a}$$

Friedmann's Second Equation: $$\ddot{a} = -\frac{4\pi G}{3}\left(\rho + \frac{3P}{c^2}\right)a, \tag{11.4b}$$

Friedmann's Third Equation: $$\dot{\rho} = -3\left(\rho + \frac{P}{c^2}\right)\frac{\dot{a}}{a}, \tag{11.4c}$$

Equation of State: $$P = P(\rho), \tag{11.4d}$$

which, as we showed in Section 7.3, can be derived without general relativity.

We have looked at three different equations of state $P = P(\rho)$:

- *Dust approximation* for matter-dominated Universe: in comoving coordinates, the matter is approximated as stationary dust particles that produce no pressure – $P = 0$.
- *Perfect-fluid approximation* for radiation-dominated Universe: the pressure induced by the movement of relativistic particles is $P = u/3 = \rho c^2/3$, where u is the energy density and ρ its mass equivalent.
- *Cosmological constant* for the dark-energy-dominated Universe: $P = -u = -\rho c^2$.

Table 11.1 Parameter w for the equations of state in different regimes: $P = w\rho c^2$.

Regime	w	Scaling with $a(t)$
Radiation-dominated	1/3	$\propto a^{-4}$
Matter-dominated	0	$\propto a^{-3}$
Dark-energy-dominated	−1	$\propto 1$

11.2. Cosmology: solutions of Friedmann's equations

To specify a cosmology, we use Friedmann's equations and choose:

1. *Curvature* of the Universe:
   - *Flat*: $k = 0$,
   - *Closed*: $k = +1$,
   - *Open*: $k = -1$;
2. *Equation of state*, generally expressed through a w parameter, defined as

$$P \equiv wc^2\rho. \tag{11.5}$$

Dominating regimes for different equations of state are given in Table 11.1 and Fig. 11.1.

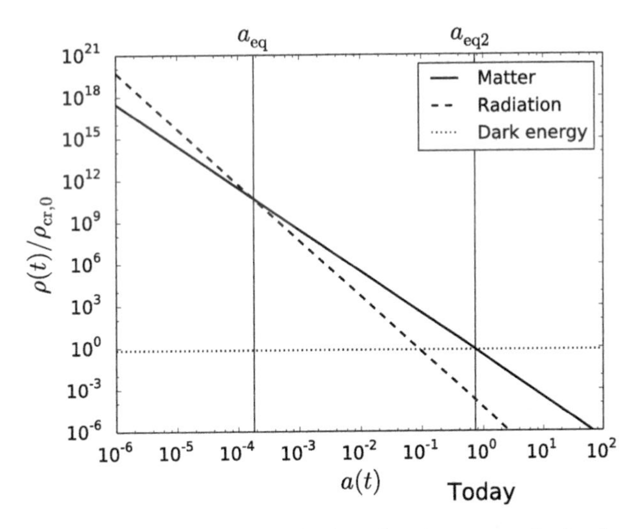

Figure 11.1 Normalized energy densities during the three epochs in the evolution of the Universe: (1) radiation-dominated $a < a_{eq}$; (2) matter-dominated $a_{eq} < a < a_{eq2}$; (3) dark energy–dominated $a > a_{eq2}$.

11.3. Expanding Universe

Solving Friedmann's equations yields a number of different cosmologies, which we derived and discussed in the earlier chapters. The most general version of Friedmann's first equation for the evolution of the scale of the Universe $a(t)$ that includes contributions from radiation, matter, dark energy, and the curvature is given by Eq. (9.14):

$$\left(\frac{\dot{a}}{a}\right)^2 = H_0^2\left[\Omega_{\mathrm{r},0}a^{-4} + \Omega_{\mathrm{m},0}a^{-3} + \Omega_{k,0}a^{-2} + \Omega_\Lambda\right]. \tag{11.6}$$

At the present time, $a = a_0 = 1$, and $H \equiv \dot{a}/a = H_0$, hence, in terms of the normalized present values of the radiation energy density, matter density, dark-energy density, and the curvature,

$$\Omega_{k,0} = 1 - \Omega_{\mathrm{r},0} - \Omega_{\mathrm{m},0} - \Omega_\Lambda \equiv 1 - \Omega_{\mathrm{T},0}. \tag{11.7}$$

In the earlier chapters, we solved Friedmann's first equation given in Eq. (11.6) by considering only the dominant component of the right-hand side of the equation, one at the time. This is a decent approximation (albeit with unphysically abrupt transitions) because of the different manner in which the energy/matter constituents of the Universe expressed on the right-hand side of the equation scale with the size of the Universe a: a component that drops off steeper (smaller slope in a) dominates earlier epochs of the evolution. Therefore according to Eq. (11.6), the order of dominance of energy/mass constituents is: radiation ($\propto a^{-4}$), matter ($\propto a^{-3}$), curvature (if $k \neq 0$; $\propto a^{-2}$), and dark energy ($\propto a^0$), as illustrated in Figs. 8.2 and 11.2. The transition between the regimes of dominance is computed by equating the contribution of the two bounding components (see Appendices E and F). This approximation of considering only the dominant energy/matter constituents of the Universe makes splicing the different epochs cumbersome. A more holistic, albeit computationally more expensive, approach is to consider all components together, by explicitly solving Eq. (11.6) above.

This leads to the equation for the time elapsed in the Universe containing all of its (possible) constituents: radiation, matter, dark energy, and curvature:

$$t = \frac{1}{H_0}\int_0^a \frac{da'}{\sqrt{\Omega_{\mathrm{r},0}a'^{-2} + \Omega_{\mathrm{m},0}a'^{-1} + \Omega_\Lambda a'^2 + \Omega_{k,0}}}. \tag{11.8}$$

This relation establishes a direct dependence of the age of the Universe and its scale $t(a)$ (it can be inverted to yield $a(t)$, as shown in Figs. 8.2 and 11.2). It can also be directly integrated to $a = a_0 = 1$ to compute the current age of the Universe t_0.

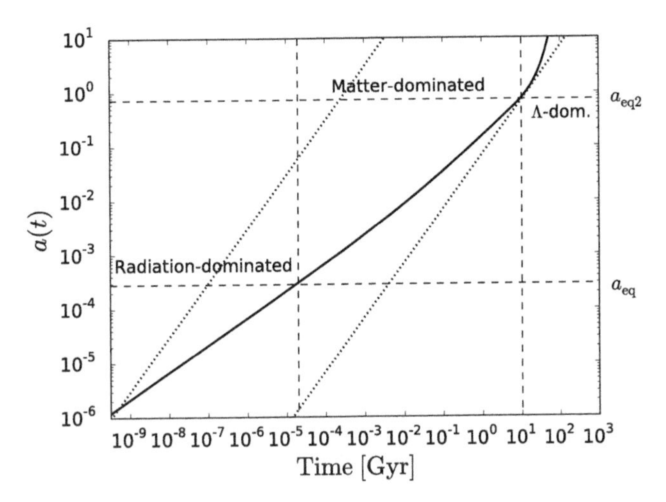

Figure 11.2 The three epochs in the history of the flat Universe: (1) radiation-dominated $a < a_{eq}$; (2) matter-dominated $a_{eq} < a < a_{eq2}$; (3) dark energy–dominated $a > a_{eq2}$. The expansion – the rate of change of $a(t)$ – during the first two epochs is sublinear (linear regime is shown in dotted lines), and the Universe is experiencing *decelerating expansion*. The expansion during the last epoch is exponential (and hence super-linear), and therefore, the rate of expansion of the Universe is increasing (*accelerating expansion*).

Sometimes, it is more convenient to express the integration in Eq. (11.8) in terms of the redshift z by making a substitution $a = 1/(1+z)$:

$$t = \frac{1}{H_0}\int_0^{\infty} \frac{dz'}{\sqrt{\Omega_{r,0}(1+z')^6 + \Omega_{m,0}(1+z')^5 + \Omega_\Lambda(1+z')^2 + \Omega_{k,0}(1+z')^4}}. \tag{11.9}$$

Of course, in both Eq. (11.8) and Eq. (11.9), the integration can be done from the present time ($a = 1$ and $z = 0$) to an event in the past, as specified by the value of the size of the Universe a or the redshift z. The result would give the time elapsed from that event.

Some of the solutions of Friedmann's equations predicted an age of the Universe that is *grossly* wrong, leading us to believe that the underlying assumptions were incorrect. The observations show that the Universe is very nearly flat, so we focus on the flat $k = 0$ cosmology. Solving for the scale factor $a(t)$ in the flat Universe – without any additional *a priori* assumptions – we obtain that the Universe is *expanding*, and that its expansion is decelerating ($\ddot{a} < 0$) during the radiation- and matter-dominated epochs, and accelerating ($\ddot{a} > 0$) during the dark-energy-dominated epoch (see Table 11.2 and

Fig. 11.2). Observations also quantify the current relative content of the Universe – how much of the critical density is found in radiation (about 0.005%), baryonic (about 4%), dark matter (about 24%), and dark energy (about 72%).

Table 11.2 Scale factor $a(t)$ for different regimes in the flat Universe. Recall from Eq. (9.27), $H = const. > 0$ in the dark-energy-dominiated regime.

Regime	$a(t)$	$\dot{a}(t)$		$\ddot{a}(t)$	
Radiation-dominated	$\propto t^{1/2}$	$\propto t^{-1/2} > 0$	expanding	$\propto -t^{3/2} < 0$	decelerating
Matter-dominated	$\propto t^{2/3}$	$\propto t^{-1/3} > 0$	expanding	$\propto -t^{4/3} < 0$	decelerating
Dark-energy-dominated	$\propto e^{Ht}$	$\propto e^{Ht} > 0$	expanding	$\propto e^{Ht} > 0$	accelerating

Exercises

1. **(Numerical)** Use your favorite plotting tool to reproduce the curves in Fig. 11.2.

CHAPTER 12

Cosmic content

It is far better to grasp the Universe as it really is than to persist in delusion, however satisfying and reassuring.

Carl Sagan

The Big Picture: In this chapter, we are going to discuss the content of the Universe: first photons and neutrinos, and their relative abundances; next, the matter content – baryonic and dark matter; and finally finish with the dark energy.

12.1. Particle distribution function

Heisenberg's Uncertainty Principle states that the position and momentum of a particle are simultaneously knowable only to within an uncertainty

$$\Delta x \Delta p_x > 2\pi\hbar, \tag{12.1}$$

where $\hbar$ is the reduced Planck constant (related to the Planck constant: $h = 2\pi\hbar$).

The same restriction holds for the other two spatial dimensions, hence,

$$\Delta x \Delta p_x \Delta y \Delta p_y \Delta z \Delta p_z > (2\pi\hbar)^3 \quad \rightarrow \quad d^3\mathbf{x} d^3\mathbf{p}. \tag{12.2}$$

The constant $(2\pi\hbar)^3$ defines the 6D volume "element" in *phase space* (space consisting of coordinates $\boldsymbol{x}$ and momenta $\boldsymbol{p}$). Therefore no particle can be localized in a phase-space volume smaller than $(2\pi\hbar)^3$, so this becomes the unit size of the phase-space.[1] This means that the number of fermions that can fit in an arbitrary volume element

[1] *Pauli's Exclusion Principle* states that two identical fermions cannot occupy the same quantum state. Therefore each fermion must occupy the unit size of the phase space $(2\pi\hbar)^3$. Any compaction in one phase-space coordinate must be compensated in expansion in others, in order to keep the phase-space volume above the minimum size of $(2\pi\hbar)^3$. If the fermions are to be packed in a small space volume dV, they must have large momenta different from those of the other fermions in the volume. Thus packing of particles in the h^3 volume elements pushes them to larger momenta, leading to larger pressure. For example, in the most compact gravitationally non-singular bodies like white dwarfs and neutron stars, it is this dense spatial packing of particles that leads to extraordinary large pressures needed to keep these bodies from gravitational collapse.

Relativity and Cosmology
https://doi.org/10.1016/B978-0-44-323542-9.00020-1

$d^3\boldsymbol{x}d^3\boldsymbol{p}$ is

$$\frac{d^3\boldsymbol{x}d^3\boldsymbol{p}}{(2\pi\hbar)^3}f(\boldsymbol{x},\boldsymbol{p}), \tag{12.3}$$

where $f(\boldsymbol{x},\boldsymbol{p})$ is a *particle distribution function.*

The total number of particles N found in some arbitrary 6D phase space volume is simply given by

$$N = \int \frac{d^3\boldsymbol{x}d^3\boldsymbol{p}}{(2\pi\hbar)^3}f(\boldsymbol{x},\boldsymbol{p}). \tag{12.4}$$

Similarly, integrating the particle distribution function over the momenta $\boldsymbol{p}$ only yields *number density n*:

$$n(\boldsymbol{x}) = \int \frac{d^3\boldsymbol{p}}{(2\pi\hbar)^3}f(\boldsymbol{x},\boldsymbol{p}). \tag{12.5}$$

Particle distribution functions allow us to compute the *total amount of some quantity over the entire distribution*: just specify the desired quantity q in terms of the phase-space coordinates $(\boldsymbol{x},\boldsymbol{p})$ for one arbitrary particle, and then integrate that quantity over the distribution function:

$$q_{\text{total}} = \int \frac{d^3\boldsymbol{x}d^3\boldsymbol{p}}{(2\pi\hbar)^3}q(\boldsymbol{x},\boldsymbol{p})f(\boldsymbol{x},\boldsymbol{p}). \tag{12.6}$$

Similarly, a mean of a quantity q over the distribution can be found by dividing the total quantity by the integration volume:

$$\bar{q} = \frac{\int \frac{d^3\boldsymbol{x}d^3\boldsymbol{p}}{(2\pi\hbar)^3}q(\boldsymbol{x},\boldsymbol{p})f(\boldsymbol{x},\boldsymbol{p})}{\int \frac{d^3\boldsymbol{x}d^3\boldsymbol{p}}{(2\pi\hbar)^3}f(\boldsymbol{x},\boldsymbol{p})} = \frac{\int \frac{d^3\boldsymbol{x}d^3\boldsymbol{p}}{(2\pi\hbar)^3}q(\boldsymbol{x},\boldsymbol{p})f(\boldsymbol{x},\boldsymbol{p})}{N}. \tag{12.7}$$

Integrating a quantity over the momenta alone $d^3\boldsymbol{p}$ yields its *space density*:

$$\rho_q(\boldsymbol{x}) = \int \frac{d^3\boldsymbol{p}}{(2\pi\hbar)^3}q(\boldsymbol{x},\boldsymbol{p})f(\boldsymbol{x},\boldsymbol{p}). \tag{12.8}$$

For isotropic systems, in which particle motion is independent of the direction, the particle distribution function depends only on the magnitude of the momenta, p, and not on their individual components (p_x, p_y, p_z). Exploiting this symmetry, we can rewrite the momentum components in spherical polar coordinates, as

$(p\cos\theta, p\sin\theta\cos\phi, p\sin\theta\sin\phi)$, so that the volume element is $p^2\sin\theta\, dp d\theta\, d\phi$. Integrating over the angles θ and ϕ, because of the distribution's independence on the angles (isotropy), yields

$$\int d^3\mathbf{p} = 4\pi p^2 \int dp \quad \rightarrow \quad d^3\mathbf{p} = 4\pi p^2 dp \tag{12.9}$$

and

$$\frac{d^3\mathbf{x} d^3\mathbf{p}}{(2\pi\hbar)^3} f(\mathbf{x}, \mathbf{p}) = \frac{4\pi\, d^3\mathbf{x}}{(2\pi\hbar)^3} p^2 f(\mathbf{x}, p). \tag{12.10}$$

12.2. Particle distribution function of species

The particle distribution functions of the different species is given by the Bose–Einstein distribution for bosons (particles with an integer spin, such as photons, W and Z bosons, gluons, mesons, etc.):

$$f_{\mathrm{BE}}(p) = \frac{1}{e^{(E(p)-\mu)/(kT)} - 1}, \tag{12.11}$$

and the Fermi–Dirac distribution for fermions (particles with a half-integer spin, such as quarks, baryons, leptons, etc.):

$$f_{\mathrm{FD}}(p) = \frac{1}{e^{(E(p)-\mu)/(kT)} + 1}, \tag{12.12}$$

where k is the Boltzmann constant, $E(p) = \sqrt{p^2c^2 + m^2c^4}$ and μ is the chemical potential, which is much smaller than the temperature T for almost all particles at almost all times, and can therefore be safely ignored in most of the calculations (see Appendix H).

These distributions are for the smooth Universe, and represent a zero-order approximation. They, therefore, do not depend on positions $\mathbf{x}$ or on the direction of the momentum $\mathbf{p}$, but only on the magnitude of the momentum p.

The properties of species specified by the distribution function $f(\mathbf{x}, \mathbf{p})$ are computed by integrating quantities over the distribution function. For example, the energy density of a species i, $u_i = \rho_i c^2$ is given by

$$u_i(\mathbf{x}) = \rho_i(\mathbf{x})c^2 = g_i \int \frac{d^3\mathbf{p}}{(2\pi\hbar)^3} f_i(\mathbf{x}, \mathbf{p}) E(p), \tag{12.13}$$

where g_i is the degeneracy of the species (for instance, $g_i = 2$ for the photon for its spin states).

Similarly, the pressure of a species i can be expressed as (the derivation is left as an exercise):

$$P_i(\mathbf{x}) = g_i \int \frac{d^3\mathbf{p}}{(2\pi\hbar)^3} f_i(\mathbf{x}, \mathbf{p}) \frac{p^2c^2}{3E(p)}. \tag{12.14}$$

12.3. Entropy density

The entropy density is defined as[a]:

$$s \equiv \frac{\rho c^2 + P}{T}. \tag{12.15}$$

[a] When chemical potential is negligible, as is almost always the case in cosmology. See Appendix H.

The unit of entropy is J/K, so the unit of the entropy density must be J/(K m^3), which is indeed what we obtain from the expression above: ρc^2 and P are energy density in units of J/m^3 and T is the temperature in K. To compute how the entropy density scales with the scale factor a, rewrite the second Friedmann equation (Eq. (7.34c)):

$$\dot{\rho} + 3\left(\rho + \frac{P}{c^2}\right)\frac{\dot{a}}{a} = 0$$

$$a^{-3}\frac{\partial}{\partial t}\left(\rho a^3\right) + 3\frac{\dot{a}}{a}\frac{P}{c^2} = 0$$

$$a^{-3}\frac{\partial}{\partial t}\left(\left(\rho + \frac{P}{c^2}\right)a^3\right) - \frac{1}{c^2}\frac{\partial P}{\partial t} = 0.$$

Combining the equation above with the result (see the exercises below)

$$\frac{\partial P}{\partial T} = \frac{\rho c^2 + P}{T} \tag{12.16}$$

and the fact that, due to the chain rule,

$$\frac{\partial P}{\partial t} = \frac{\partial P}{\partial T}\frac{\partial T}{\partial t}, \tag{12.17}$$

we obtain

$$a^{-3}\frac{\partial}{\partial t}\left(\left(\rho + \frac{P}{c^2}\right)a^3\right) - \frac{1}{c^2}\frac{\partial T}{\partial t}\frac{\rho c^2 + P}{T} = \frac{1}{c^2}a^{-3}T\frac{\partial}{\partial t}\left(\frac{(\rho c^2 + P)a^3}{T}\right) = 0. \tag{12.18}$$

The quantity in brackets then must be constant:

$$\frac{(\rho c^2 + P)a^3}{T} = sa^3 = const.$$

Therefore the entropy density scales with the scale factor as $s \propto a^{-3}$.

12.4. Cosmic microwave background photons

The very early Universe (discussed in more detail in Chapter 13) was a hot, dense "soup" of photons, electrons, and baryons. The photons could not travel very far at all before being absorbed and quickly reemitted by the electrons, *i.e.*, their *mean-free path*, the average distance between interactions, was miniscule. This vanishing mean-free path meant that the Universe was opaque. As the Universe expanded and cooled, the number density of the particles, and therefore the incidence of these interactions, slowed and stopped entirely by about 380,000 years after the Big Bang. At that time, the photons decoupled from matter and became free to travel with virtually infinite mean-free paths. These photons, which innundate the Universe today, are called *cosmic microwave background* (CMB) photons.

Before decoupling, CMB photons were in equilibrium with the rest of the Universe, and therefore shared a common temperature. After decoupling, matter and CMB photons evolved separately. At any one time t_1 thereafter, CMB photons continued to have the same[2] temperature $T_\gamma(t_1)$, even as it decreased over time.

The energy density of CMB photons can be found by using Eq. (12.13) with the Bose–Einstein distribution given in Eq. (12.11):

$$u_\gamma = g_\gamma \int \frac{d^3\mathbf{p}}{(2\pi\hbar)^3} \frac{E(p)}{e^{E/(kT_\gamma)} - 1} = 2 \int \frac{d^3\mathbf{p}}{(2\pi\hbar)^3} \frac{cp}{e^{cp/(kT_\gamma)} - 1}, \tag{12.19}$$

where we have used $g_\gamma = 2$, $E(p) = \sqrt{p^2c^2 + m^2c^4} = cp$ for massless photons, and neglected the chemical potential μ. Using $d^3\mathbf{p} = 4\pi p^2 dp$, and making a substitution $x = cp/(kT_\gamma)$ yields

$$u_\gamma = \frac{8\pi}{(2\pi\hbar)^3} \int_0^\infty \frac{cp^3}{e^{cp/(kT_\gamma)} - 1} dp = \frac{8\pi}{(2\pi\hbar)^3} \frac{k^4 T_\gamma^4}{c^3} \int_0^\infty \frac{x^3}{e^x - 1} dx = \frac{8\pi}{(2\pi\hbar)^3} \frac{k^4 T_\gamma^4}{c^3} \frac{\pi^4}{15}. \tag{12.20}$$

Therefore we find that the energy density of the CMB photons is

$$u_\gamma = a_r T_\gamma^4, \tag{12.21}$$

[2] To within about one part in 100,000, as detailed in Chapter 14.

where

$$a_r \equiv \frac{\pi^2 k^4}{15\hbar^3 c^3}, \tag{12.22}$$

is the radiation constant. Here, we also used the result

$$\int_0^\infty \frac{x^3}{e^x - 1} dx = \frac{\pi^4}{15}. \tag{12.23}$$

We have derived earlier that the energy density of radiation scales as $\rho_\gamma \propto a^{-4}$ (see Eq. (5.43)). Since, from Eq. (12.21), $\rho_\gamma \propto T_\gamma^4$, it follows that $T_\gamma \propto a^{-1}$ and $T_\gamma a = T_{\gamma,0} a_0 = const.$

Therefore the temperature of the CMB photons scales with the scale factor as

$$T_\gamma = T_{\gamma,0} a^{-1}, \tag{12.24}$$

where $T_{\gamma,0} = 2.73$ K is the temperature of the CMB measured today.

In terms of the critical density $\rho_{cr,0}$, we have

$$\Omega_\gamma \equiv \frac{u_\gamma}{c^2 \rho_{cr,0}} = \frac{a_r T_\gamma^4}{c^2 \rho_{cr,0}} = \frac{a_r \left(T_{\gamma,0} a^{-1}\right)^4}{c^2 \rho_{cr,0}} = \frac{a_r T_{\gamma,0}^4}{c^2 \rho_{cr,0} a^4}, \tag{12.25}$$

where c^2 in the denominator is because $\rho_{cr,0}$ is the mass density, and $c^2 \rho_{cr,0}$ is the energy density. From Eq. (8.10), $\rho_{cr,0} = 1.87 \times 10^{-29} h^2$ g cm^{-3}, hence,

$$\Omega_\gamma = \frac{u_\gamma}{c^2 \rho_{cr,0}} = \frac{2.48 \times 10^{-5}}{h^2 a^4}. \tag{12.26}$$

If we take $h \approx 0.73$, then the fractional content of the Universe due to CMB radiation *today* is

$$\Omega_{\gamma,0} \equiv \frac{u_{\gamma,0}}{c^2 \rho_{cr,0}} = 4.65 \times 10^{-5}. \tag{12.27}$$

12.5. Cosmic neutrino background

Similarly to the CMB photons, *the cosmic neutrinos* were in equilibrium with the rest of the very early Universe. However, owing to their exceedingly small *interaction cross section*,[3] they decoupled from the rest of the Universe much earlier than the CMB photons.

In order to compute the relative energy density of *the cosmic neutrino background* (CNB), we need to relate its temperature to the temperature of the CMB photons.

Before the decoupling of neutrinos, at some a_1, the plasma has a uniform temperature of, say, T_1. The pressure due to CMB radiation (photons) is given by the perfect-fluid approximation as

$$P_\gamma = \frac{1}{3}u_\gamma, \tag{12.28}$$

so the contribution to the entropy for each spin state is (recall Eq. (12.21) has a factor $g_\gamma = 2$ reflecting 2 spin states)

$$s_\gamma = \frac{u_\gamma + P_\gamma}{T_1} = \frac{4u_\gamma}{3T_1} = \frac{4\,(a_r T_1)^4}{3T_1} = \frac{4}{3}a_r^4 T_1^3. \tag{12.29}$$

Photons are bosons, and hence subject to Bose–Einstein statistics, which, as we saw in Eq. (12.23) leads to the integral

$$I_{\rm BE} \equiv \int_0^\infty \frac{x^3}{e^x - 1}dx = \frac{\pi^4}{15}. \tag{12.30}$$

Neutrinos are leptons, and hence fermions, so they are subject to the Fermi–Dirac distribution, which leads to the integral

$$I_{\rm FD} \equiv \int_0^\infty \frac{x^3}{e^x + 1}dx = \frac{7\pi^4}{120} = \frac{7}{8}I_{\rm BE}. \tag{12.31}$$

Therefore the contribution of neutrinos will be 7/8 of the contribution of photons. Just before the electron–positron annihilation, there are the following fermions: electrons (2 spin states), positrons (2 spin states), neutrinos (3 generations and 1 spin state), and anti-neutrinos (3 generations and 1 spin state), and the following bosons: photons (2 spin states). Therefore before the electron–positron annihilation (at some $a = a_1$), the entropy density is given by the sum of all entropies of species:

$$s(a_1) = s_\gamma\left(2 + \frac{7}{8}(2+2+3+3)\right) = \frac{43}{4}s_\gamma = \frac{43}{3}a_r^4 T_1^3. \tag{12.32}$$

After the electron–positron annihilation, the temperatures of photons and neutrinos are no longer equal. Neutrinos decoupled slightly before the annihilation.

[3] Roughly speaking, the interaction cross section σ measures how close another particle has to come in order to cause an interaction. The rate of interactions scales linearly with both the number densities of the interacting particles and the interaction cross section. Therefore the expanding Universe rapidly decreases the number densities of interacting species and, consequently, the rates of their interaction.

The temperature of the neutrino background evolves with the scale factor as $T_\nu = T_{\nu,0}a^{-1}$ (just like for the photon background).

Photons were still coupled to the plasma during the electron–positron annihilation. As the electrons and positrons annihilate, they produce high-energy photons that quickly reach equilibrium with the rest of the plasma. This effectively raises the equilibrium temperature of the plasma, including photons, T_γ. The entropy density after the annihilation (at some $a = a_2$) is therefore

$$s(a_2) = 2\frac{4}{3}a_r^4T_\gamma^3 + 6\frac{7}{8}\frac{4}{3}a_r^4T_\nu^3 = \frac{8}{3}a_r^4T_\gamma^3 + 7a_r^4T_\nu^3 = \frac{1}{3}a_r^4\left(8T_\gamma^3 + 2T_\nu^3\right). \tag{12.33}$$

However, entropy density s scales as a^{-3}, so

$$sa^3 = s(a_1)a_1^3 = s(a_2)a_2^3 = const., \tag{12.34}$$

which leads to

$$\frac{43}{3}a_r^4T_1^3a_1^3 = \frac{1}{3}a_r^4\left(8T_\gamma^3 + 21T_\nu^3\right)a_2^3. \tag{12.35}$$

Neutrino temperature scales throughout as a^{-1}:

$$Ta = T_1a_1 = T_\nu a_2 = const., \tag{12.36}$$

hence, equating the two entropies before and after, given in Eq. (12.32) and Eq. (12.33),

$$\begin{aligned}
&\frac{43}{3}a_r^4T_1^3a_1^3 = \frac{43}{3}a_r^4T_\nu^3a_2^3 = \frac{1}{3}a_r^4\left(8\left(\frac{T_\gamma}{T_\nu}\right)^3 + 21\right)T_\nu^3a_2^3,\\
&\rightarrow \quad \frac{43}{8} = \left(\frac{T_\gamma}{T_\nu}\right)^3 + \frac{21}{8} \quad \rightarrow \quad \left(\frac{T_\gamma}{T_\nu}\right)^3 = \frac{22}{8}\\
&\rightarrow \quad \frac{T_\gamma}{T_\nu} = \left(\frac{11}{4}\right)^{1/3} \approx 1.4, \quad \text{or} \quad \frac{T_\nu}{T_\gamma} = \left(\frac{4}{11}\right)^{1/3} \approx 0.71.
\end{aligned} \tag{12.37}$$

This means that the cosmic neutrino temperature is lower by about a factor $(4/11)^{1/3}$ (about 29%) than the CMB radiation (photon) temperature. This difference comes from the heating of the photons by the annihilation of electrons and positrons. Given that $T_{\gamma,0} \approx 2.73$ K at the present time, one expects that $T_{\nu,0} \approx 1.95$ K.[a]

[a] CNB has not been detected yet due to a tiny neutrino interaction cross section.

Now that we can relate the temperature of neutrinos T_ν to the temperature of photons T_γ, we can compute the energy density of the neutrinos (which are fermions, and hence subject to the Fermi–Dirac distribution function):

$$u_\nu = g_\nu \int \frac{d^3\mathbf{p}}{(2\pi\hbar)^3} \frac{E(p)}{e^{E/(kT_\nu)}+1} = 6\int \frac{d^3\mathbf{p}}{(2\pi\hbar)^3} \frac{cp}{e^{cp/(kT_\nu)}+1}, \tag{12.38}$$

where $g_\nu = 6$ (3 flavors with 2 distinct particle types: ν_e, ν_μ, ν_τ, $\bar{\nu}_e$, $\bar{\nu}_\mu$, $\bar{\nu}_\tau$), $E(p) = \sqrt{c^2p^2+m^2c^4} = p$ for neutrinos, and we neglected the chemical potential μ. Again, after noting that $d^3p = 4\pi p^2 dp$, and making a substitution $x = cp/(kT_\nu)$,

$$\begin{aligned} u_\nu &= \frac{24\pi}{(2\pi\hbar)^3}\int_0^\infty \frac{cp^3}{e^{cp/(kT_\nu)}+1}dp = \frac{24\pi}{(2\pi\hbar)^3}\frac{k^4T_\nu^4}{c^3}\int_0^\infty \frac{x^3}{e^x+1}dx \\ &= \frac{24\pi}{(2\pi\hbar)^3}\frac{k^4T_\nu^4}{c^3}I_{\mathrm{FD}} = \frac{3\pi}{\pi^3}\frac{k^4T_\nu^4}{\hbar^3c^3}\frac{7}{8}\frac{\pi^4}{15} = \frac{21}{8}\frac{\pi^2k^4}{15\hbar^3c^3}T_\nu^4 \\ \rightarrow \quad u_\nu &= \frac{21}{8}a_rT_\nu^4 = \frac{21}{8}\left(\frac{4}{11}\right)^{4/3}a_rT_\gamma^4, \end{aligned} \tag{12.39}$$

where we used Eq. (12.37) in the last step. In terms of the energy density of photons,

$$u_\nu = \frac{21}{8}\left(\frac{4}{11}\right)^{4/3}\rho_\gamma. \tag{12.40}$$

We also have

$$\Omega_\nu = \frac{21}{8}\left(\frac{4}{11}\right)^{4/3}\Omega_\gamma = \frac{21}{8}\left(\frac{4}{11}\right)^{4/3}\frac{2.48\times10^{-5}}{h^2a^4} = \frac{1.69\times10^{-5}}{h^2a^4}, \tag{12.41}$$

so that the ratio of the neutrino density to the critical density *today* is (if we take $h \approx 0.73$)

$$\Omega_{\nu,0} \equiv \frac{u_{\nu,0}}{c^2\rho_{\mathrm{cr},0}} = 3.17\times10^{-5}. \tag{12.42}$$

All of the calculations above were done assuming that the neutrinos are massless. However, observations of solar neutrinos indicate that they change flavors on their way from the Sun to us, which can only happen if they have mass. The observations of atmospheric neutrinos suggest that at least one neutrino has mass larger than 0.05 eV. In that case, for a massive neutrino, $E(p) = \sqrt{c^2p^2+m_\nu^2c^4} \neq cp$, so the integral in Eq. (12.39) becomes (with $g_\nu = 2$ for one flavor of neutrinos with 2 spin states)

$$u_\nu = \frac{8\pi}{(2\pi\hbar)^3}\int_0^\infty \frac{p^2\sqrt{c^2p^2+m_\nu^2c^4}}{e^{\sqrt{c^2p^2+m_\nu^2c^4}/(kT_\nu)}+1}dp. \tag{12.43}$$

12.6. Baryonic matter

When using the term "baryonic matter," both baryons and electrons are implied. Electrons are not baryons – they are leptons – but given that the mass of an electron is nearly 2000 times smaller than the mass of a proton or a neutron, electron contribution is negligible.

Unlike the energy density of CMB radiation, which can be described as a gas with a temperature and vanishing chemical potential (perfect-fluid approximation), the baryonic density must be directly measured (dust approximation). The different methods that measure baryonic density at varying redshifts z largely agree that it is about 2–5% of the critical density *today*:

$$\Omega_{b,0} \equiv \frac{\rho_{b,0}}{\rho_{cr,0}} = 0.02 - 0.05. \tag{12.44}$$

We expect that the total amount of baryonic matter is constant, so with the expanding Universe, the fractional energy density scales as $\rho_b \propto a^{-3}$, hence,

$$\Omega_b = \frac{\rho_b}{\rho_{cr,0}} = \frac{\rho_{b,0}}{\rho_{cr,0}} a^{-3} = \Omega_{b,0} a^{-3}. \tag{12.45}$$

Several methods are used to gauge the baryon content of the Universe:

1. Directly observing visible matter in galaxies. It has been found that the largest contribution comes from the gas in galaxy clusters, while stars in galaxies account for only a comparatively small fraction. This approach estimates $\Omega_{b,0} \approx 0.02$.
2. Looking at spectra of distant galaxies and measuring the light absorption quantifies the amount of hydrogen the light encounters along the way. Baryon density is then inferred from this estimate. This approach roughly estimates $\Omega_{b,0} h^{1.5} \approx 0.02$ [23].
3. Computing the baryon content of the Universe from the anisotropies of the CMB radiation. This approach puts fairly stringent limits on the baryon content at about $\Omega_{b,0} h^2 = 0.024^{+0.004}_{-0.003}$.
4. Inferring the baryon content of the Universe form the light-element abundances. These pin down the baryon content to $\Omega_{b,0} h^2 = 0.0205 \pm 0.0018$.

These estimates are in fairly good agreement. They put a rough baryonic content of the Universe at about 2–5% of the critical density. However, as we shall soon see, the total matter density in the Universe is significantly higher than that, so there must be another form of matter other than baryonic.

12.7. Dark matter

The first evidence for the *dark matter* was provided by a Swiss astrophysicist Fritz Zwicky [24] in 1933. He used the virial theorem to show that the observed (luminous) matter was not nearly enough to keep the Coma cluster of galaxies together.

In 1939, Horace Babcock [25] reported on his observations that the rotation curves for Andromeda suggested that the mass-to-light ratio increased radially. However, he did not directly recognize this as an argument for the existence of some "missing mass," but explained it away as the absorption of light or to modified dynamics in the outer portions of the disk.

For a couple of decades, the "missing mass problem" was largely ignored, until Vera Rubin and Kent Ford [26] in the late 1960s and early 1970s measured velocity curves of edge-on spiral galaxies, starting again with Andromeda, to a theretofore unprecedented accuracy. Their work unequivocally demonstrated that most stars in spiral galaxies orbit the center at roughly the same speed, which suggested the presence of some missing mass whose contribution in the outer parts dwarfs that of the luminous mass. This was consistent with the spiral galaxies being embedded in a much larger halo of invisible mass ("dark-matter halo").

One of the oldest and most straightforward methods for estimating the matter density of the Universe is the mass-to-light ratio technique. The average ratio of the observed mass to light of the largest possible system is used; assuming that the sample is fair, it can be multiplied by the total luminosity density of the Universe to obtain the total mass density ρ_m. Zwicky was the first to do this with the Coma cluster, but many followed.

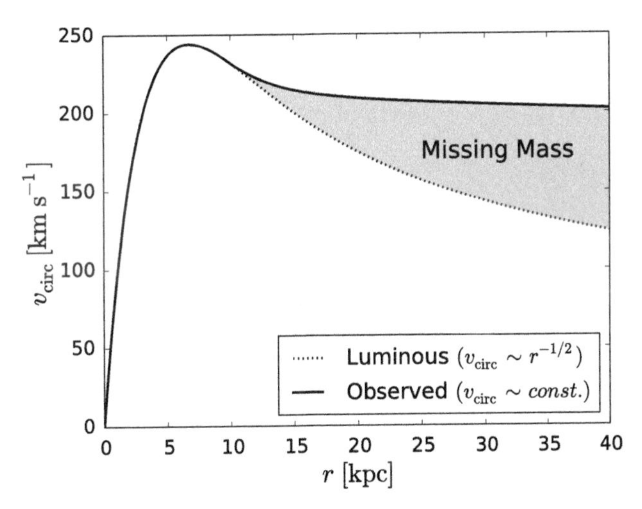

Figure 12.1 A spiral galaxy's rotation curve – its circular velocity v_{circ} as a function of radius r. The dotted curve shows the result of computing v_{circ} from Eq. (12.46) where only luminous matter is included. It exhibits a $\sim r^{-1/2}$ falloff. The solid line depicts the observed rotation curves. The discrepancy between the two curves (shaded region) – the "missing mass" – is commonly attributed to dark matter.

12.7.1 Evidence for dark matter: rotation curves in disk galaxies

Consider a star with mass $m_\star$ and rotational velocity $v_r(r)$ orbiting around a galactic center on a circular orbit at some radius r. Its motion balances the outward-acting centrifugal force and the inward-acting gravitational attraction of the mass enclosed within a sphere traced out by its orbit:

$$\frac{m_\star v_r^2}{r} = G\frac{m_\star M(r)}{r^2} \quad \rightarrow \quad v_r = \sqrt{\frac{GM(r)}{r}}, \tag{12.46}$$

where

$$M(r) = 4\pi \int_0^r \rho_l(r') r'^2 dr', \tag{12.47}$$

is the galaxy mass enclosed within the sphere of radius r.

If the luminous matter ρ_l was all there was, in the outer parts of the galaxy (as $r \rightarrow \infty$), the total enclosed mass in Eq. (12.47) would approach a constant: $M(r) \rightarrow const.$ That would, according to Eq. (12.46), lead to stellar rotation velocity curves dropping off as

$$v_r(r) \propto r^{-1/2}. \tag{12.48}$$

Kinematic observations of disk galaxies at different radii give us the true rotation curves, the *measured* velocity of stars $v_r(r)$ as a function of radius. It is observed that the rotation curves are actually flat out to an order of magnitude beyond the galactic disk scale length[4] (illustrated in Fig. 12.1). Observed flat rotation curves at large radii imply that, according to Eq. (12.46),

$$v_\infty \equiv \lim_{r\rightarrow\infty} v_r(r) = \frac{GM(r)}{r} = const. \quad \rightarrow \quad \lim_{r\rightarrow\infty} M(r) \sim r. \tag{12.49}$$

v_∞ is the constant value of the flat rotation curve. Using Eq. (12.47), we find that

$$M(r) = 4\pi \int_0^r \rho(r') r'^2 dr' = r \quad \rightarrow \quad \rho(r) \sim r^{-2}. \tag{12.50}$$

The outer parts of the galaxy are dominated by some invisible matter, and that matter density scales as r^{-2}. This matter has been dubbed *dark matter.* The structure containing this dark matter that envelops disk galaxies is called the *dark-matter halo.*

[4] The measurements of rotation curves is usually based on Doppler shift of gas emissions. The gas in galaxies extends well beyond the stars.

12.7.2 Other evidence for dark matter

There are other methods that independently suggest the existence and quantify the amount of dark matter in the Universe. They include:

- *Gravitational lensing* is a direct consequence of general relativity[5]: the trajectory of a photon is affected by the curvature of spacetime induced by the presence of a massive object (lens). There are three categories of gravitational lensing:
  - *Microlensing*: Small-scale gravitational effect in which the lensing mass is too low (mass of a planet or a star) for the deflection of light to be easily detected, but the apparent brightening of the source may still be observed. The timescale over which microlensing takes place is on the order of seconds to years. With the changing alignment of the source, the lensing mass, and the observer, the apparent brightness and deflection will also change.
  - *Weak gravitational lensing*: Small distortions in the shapes of background galaxies can be created via weak lensing by foreground galaxy clusters. Statistical averaging of these small distortions yields a mass estimate of the cluster.
  - *Strong gravitational lensing*: Light rays leaving a source in different directions are nearly focused on the same spot (the observer here on Earth) by the intervening galaxy or cluster of galaxies. It produces multiple distorted images of the source from which the mass and shape of the lens can be inferred.
- The baryons-to-matter ratio in clusters of galaxies, which are the largest known virialized structures (in equilibrium), are likely representative of the Universe as a whole. If a good estimate of the baryonic matter Ω_b is adopted from the previously described methods, measuring the baryons-to-matter ratio $f_b \equiv \Omega_b/\Omega_m$ in these clusters will yield the estimate of the fractional density of matter Ω_m. The visible (baryonic) matter in clusters of galaxies is largely in hot ionized intracluster gas, with only a small, negligible fraction in stars (about an order of magnitude smaller). This means that the ratio f_b is well approximated by the ratio of gas-to-matter f_g, which can be measured via:
  - *X-ray spectrum*: Measure the mean gas temperature from the overall shape of the X-ray spectrum, and the absolute value of the gas density from the X-ray luminosity.
  - *Sunyaev–Zel'dovich effect* [27]: As the CMB radiation passes through the cluster whose baryonic mass is dominated by gaseous ionized intracluster medium (ICM), a fraction of photons inverse-Compton scatter off the hot electrons of the ICM. The intensity of the CMB radiation is therefore boosted in energy as compared to the unscattered CMB. This boost is in magnitude proportional to the number of scatterers, which, in turn, depend on the size of the electron cloud and its density.

[5] The first observation of gravitational lensing provided the first and the most notable confirmation of general relativity: the solar eclipse of 1919 confirmed that the Sun bends light that passes near it.

- The Bullet cluster was created following the collision of two large clusters of galaxies. As the two clusters collided, the hot gas in each cluster was slowed by a drag force. Dark matter, however, was not slowed down by the collision, since it only interacts through gravity. Consequently, during the collision, dark matter from the two clusters moved ahead of the hot gas, producing the separation of the dark and normal matter seen in the observations. If hot gas had been the most massive component in the clusters, as posited by alternative theories of gravity, such separation would not occur. The Bullet cluster ision shows that the presence of dark matter is required [28].
- Dark matter plays an important role in structure formation because it interacts with its surrounding only through gravity. Its gravitational collapse via the Jeans instability is therefore not opposed by any other force, such as radiation pressure (as is the case with ordinary matter). Therefore dark matter starts to collapse into dark-matter halos well before the ordinary matter. Without dark matter, galaxy formation would occur considerably later than observed.
- Anisotropies in the CMB radiation, which will be discussed in Chapter 14.

These independent methods, along with others not mentioned here, provide a compelling body of evidence that the baryon density is of the order of $\sim 5\%$ of the critical density, while the total matter density is about five times larger. This clearly implies that most of the matter in the Universe must not be baryons. It must be in some other form – *dark matter*.

From the standpoint of cosmology, the curvature of the Universe and the cosmic inventory, dark matter is treated on an equal footing with baryonic matter – it scales with the expanding Universe as $\rho_{\rm dm} \propto a^{-3}$ and contributes to the total energy density budget of the Universe.

12.7.3 Dark-matter candidates

Baryonic dark matter: MAssive Compact Halo Objects (MACHOs)

The initial mass function. Our ability to observe stars has limitations – it cuts off at some lower-level luminosity. The mass distribution of stars as set during the process of star formation – *initial mass function* (IMF) – is roughly approximated by

$$\frac{dn}{dm} \propto m^{-\alpha}, \tag{12.51}$$

with $\alpha \approx 2.35$ [29]. This and similar models are motivated empirically. We obtain the total density due to stars down to some lowest observable stellar mass m_c by integrating

$$\rho_s = \int_{m_c}^{\infty} m \; dn. \tag{12.52}$$

For the mass distribution in Eq. (12.51), the total mass density due to stars is

$$\rho_s \propto \int_{m_c}^{\infty} m^{1-\alpha}\, dm = \left. \frac{m^{2-\alpha}}{2-\alpha} \right|_{m_c}^{\infty} = \frac{m_c^{2-\alpha}}{\alpha - 2}, \tag{12.53}$$

which means that the reduction of the lower threshold of detectable stellar mass by a factor of 2 results in a stellar mass density increase of $0.5^{2-2.35}/(2.35-2) = 3.64$. More recent studies (see, for example [30,31]) have shown that the IMF flattens out to (the slope approaches $\alpha = 0$) below one solar mass ($m_c < M_\odot$). The uncertainties in the sub-stellar region – values of m_s lower than the mass necessary to maintain hydrogen-burning nuclear-fusion reactions in the cores characteristic of stars – are quite large, leading to our inability to accurately estimate the associated baryonic mass.

Brown dwarfs. Stars evolve from self-gravitating clouds of gas. Gravitational collapse of gas will cause the temperature to rise until nuclear burning can begin (a star is born!). The only way that self-gravitating gas does not yield a star is if *electron degeneracy* stops the collapse. Electron degeneracy is a consequence of the Pauli exclusion principle: no two fermions (in this case electrons) confined within a given region (in this case a star) can have the same momentum and spin. Most of the electrons in dense matter must be in a state of continual motion that results in a pressure that increases as the matter density increases. The condition for the onset of degeneracy is that the interparticle spacing becomes small enough for the uncertainty principle to become important:

$$p \leq \hbar n^{1/3}, \tag{12.54}$$

where n is the electron number density and p is the momentum. We can crudely estimate the condition for this to occur by assuming that the body is of uniform density and temperature. For a given mass, this yields an estimate of the maximum temperature that can be attained before degeneracy becomes important:

$$T_{\rm max} \approx 6 \times 10^8 \left(\frac{M_{\rm min}}{M_\odot} \right) \,{\rm K}, \tag{12.55}$$

where $M_\odot$ is the Solar mass. Hydrogen fusion requires $T \approx 10^7$ K, so the resulting minimum stellar mass is about

$$M_{\rm min} \approx 0.05\, M_\odot. \tag{12.56}$$

More accurate calculations lead to a prediction of $M_{\rm min} \approx 0.08 \pm 0.01\, M_\odot$. Objects much less massive than this will not lead to hydrogen fusion, and will therefore be virtually invisible. Such "failed stars" are called *brown dwarfs*. These objects are very difficult to detect: their spectra are heavily affected by broad molecular absorption bands, which are very difficult to model.

White dwarfs. White dwarfs form from the collapse of stellar cores once nuclear burning has ceased there. They arise when the mass of the core remnant after the death of the star is smaller than the *Chandrasekhar mass* [32], the maximum mass of a stable white dwarf star, of about 1.4 $M_{\odot}$. The end of nuclear burning in these smaller stars is followed by a *helium flash* that blows off the outer parts of the star, thus creating a planetary nebula. The remaining core contracts under its own gravity until, having reached a size similar to that of the Earth, it becomes so dense ($\sim 5 \times 10^8$ kg/m^3) that it is supported against further collapse only by the pressure of electron degeneracy. They gradually cool, becoming fainter and redder. The low luminosity of the white dwarfs (typically 10^{-3} to 10^{-4} of the Sun's) make them very difficult to detect.

Neutron stars. Neutron stars form from the collapse of stellar cores once nuclear burning has ceased there. They arise when the mass of the core remnant after the death of the star is larger than the Chandrasekhar mass of about 1.4 $M_{\odot}$ but still smaller than about $5 - 6$ $M_{\odot}$ Stars that are created from core remnants more massive than this limit of $5 - 6$ $M_{\odot}$ collapse further into a black hole.[6] The total gravitational collapse is halted by the neutron (and in smaller part proton) degeneracy. A typical neutron star, with a mass about 1.4 $M_{\odot}$, has a radius of only about 11 km, thereby giving it a density exceeding 10^{17} kg/m^3.

Black holes. When the mass of the core remnant after the death of the star is larger than about $5 - 6$ $M_{\odot}$, black holes are formed. This occurs when neutron degeneracy becomes insufficient to support the star from a total gravitational collapse, and its radius shrinks to below the Schwarzschild radius (see Section 6.3).

All of these compact (sub)stellar objects are examples of MACHOs, objects that are either very difficult or impossible to see directly. Therefore they are a form of dark matter. However, even with their contributions added to the visible baryonic matter, the total content of matter in the Universe is still significantly short.

The best limit on the possible contribution of these MACHOs comes from gravitational microlensing results from our own galaxy [33]. It is estimated that MACHOs can account for up to a few percent of the missing mass in the Milky Way.

Non-Baryonic dark matter: Weakly Interacting Massive Particles (WIMPs)

Although we presented a compelling argument that at least a small portion of the dark matter content is baryonic, there exists strong cosmological evidence that the dark matter consists of weakly interacting relic particles. The strongest case is built by primordial nucleosynthesis, or *Big Bang Nucleosynthesis* (BBN), which estimates that a baryonic contribution to the total energy density is $\Omega_{b,0} \approx 0.0125h^{-2}$. This is the contribution of the

[6] The cutoff between neutron stars and black holes is not as well established as the Chandrasekhar mass that delineates white dwarfs and neutron stars.

protons and neutrons that interacted to fix the light-element abundances at $t \approx 1$ min or so.

At the time of BBN, the Universe consisted of baryons (plus electrons, which are implicitly included in the "baryonic matter"), photons, and three species of neutrinos. To account for dark matter, one can proceed in two ways from there: (1) neutrinos have mass; or (2) there must exist some additional particle species that is a frozen-out relic from an earlier epoch.

A small neutrino mass would not affect the BBN, since the neutrinos are ultrarelativistic prior to matter–radiation equality. Other relic particles would have to be either very rare or extremely weakly coupled (even more weakly than neutrinos) in order not to effect the BBN. Either alternative would produce the *collisionless* dark matter, which is the main argument in favor of non-baryonic dark matter: the clustering power spectrum appears to be free of oscillatory features expected from the gravitational growth of perturbations in matter that is able to support sound waves.

There are a number of different WIMPs candidates for the dark-matter particle. They are called "weakly interacting" because they interact only by weak interaction and gravity, and are therefore notoriously difficult to detect.

- **Massive neutrinos**. The most obvious species of non-baryonic dark matter to consider as a dark-matter particle candidate is a massive neutrino. Because neutrinos are very weakly interacting, it is still unclear what the mass of neutrinos may be. Recent experiments only put constraints on the difference of squares of masses of two flavors of neutrinos.
- **Supersymmetric particles**. Particles that are part of supersymmetry (SUSY), and that are yet to be detected are also considered as dark-matter candidates.

There is another categorization of WIMPs, which is more descriptive of their nature: hot and cold dark matter.

Hot dark matter (HDM). HDM particles – neutrinos or any light fermion that interacts only weakly – decouple when they are relativistic, and have a number density roughly equal to that of photons. These low-mass relics are hot in the sense of possessing large quasi-thermal velocities. These velocities were larger at high redshifts, which resulted in major effects on the development of self-gravitating structures. *The predictions of HDM matter strongly disagree with observations.*

Axions. Axions (and their lighter cousins, axion-like particles, or ALPs) are hypothetical particles that are very light, yet very abundant. They were first postulated to solve a problem in quantum chromodynamics, but emerged as dark-matter candidates as they were never coupled with other particle species.

Cold dark matter (CDM). CDM particles decouple when they are non-relativistic. Most cosmologists favor the CDM theory as a description of how the Universe went from a nearly smooth initial state at early times (as demonstrated by the nearly smooth

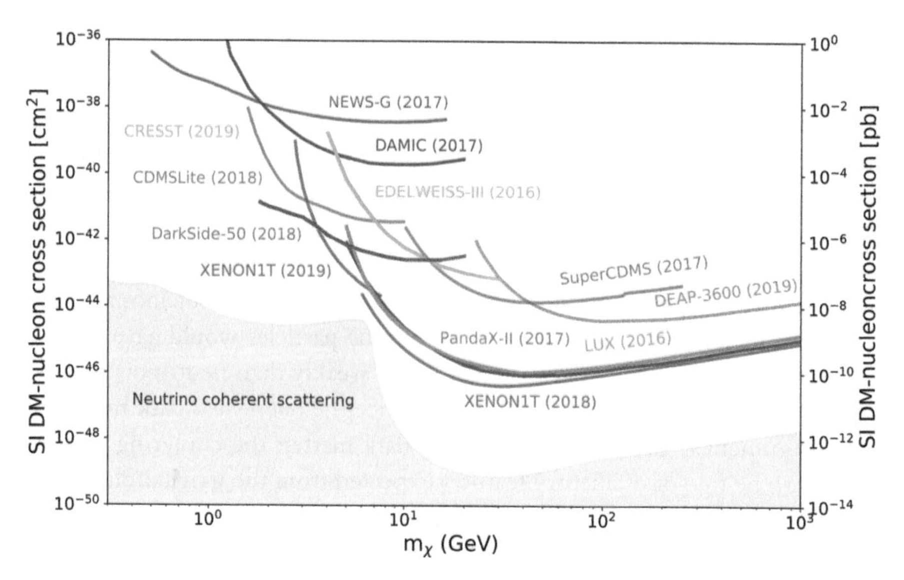

Figure 12.2 Limits on dark-matter particle's interaction cross section as a function of its mass, as established by various experiments. Adopted from the Particle Data Group [34].

CMB radiation) to the lumpy distribution of galaxies and clusters of galaxies that we observe today.

Numerous and prolonged searches for dark-matter particles are yet to turn up a direct detection. However, these searches managed to reduce the search space of intrinsic properties of such an elusive particle. Simply put, if the sensitivity of an experiment is such that it should detect dark-matter particles with certain properties, but it does not, its existence is ruled out, and the search space is reduced. Fig. 12.2 illustrates the ranges of dark-matter particle's properties, mass and interaction cross section, ruled out by various experiments.

12.8. Dark energy

The notion of a "cosmological constant" has been floating around since the time of Newton [35]. However, it is only recently that it has obtained a firm footing with theoretical and observational evidence. There are two sets of evidence that support the existence of additional energy density – "dark energy" – due to the cosmological constant:

1. **Budgetary shortfall**. The total energy density of the Universe is very close to critical, as suggested both (1) *theoretically* from the inflation in the early Universe and (2) *observationally* from the anisotropies of the CMB radiation. However, the observations can only account for about a quarter of the total critical energy density.

The remaining, unaccounted, nearly three-quarters of the density in the Universe seems to be in some smooth, unclustered form – *dark energy*.

2. **Theoretical distance–redshift relations.** Given the energy composition of the Universe (their present values: $\Omega_{\Lambda,0}$ $\Omega_{m,0}$ and $\Omega_{r,0}$), the luminosity distance $d_L(\Omega_{\Lambda,0}, \Omega_{m,0}, \Omega_{r,0})$ versus the redshift z can be theoretically computed. One can then optimize the fit of such a function to observations.
The shape of the plots of the luminosity distance d_L versus the redshift z is different in a matter-dominated Universe ($\Omega_\Lambda = 0$) from those in a dark energy–dominated Universe ($\Omega_\Lambda \gg 0$). As shown in Fig. 10.2, the luminosity distance is larger for objects at higher redshifts in a dark energy–dominated Universe. This means that objects of fixed intrinsic (absolute) brightness (*standard candles*) will appear dimmer in a Universe composed predominantly of dark energy. This concept has been so pivotal in our understanding of the composition of the Universe that we describe it in more detail in the next section.

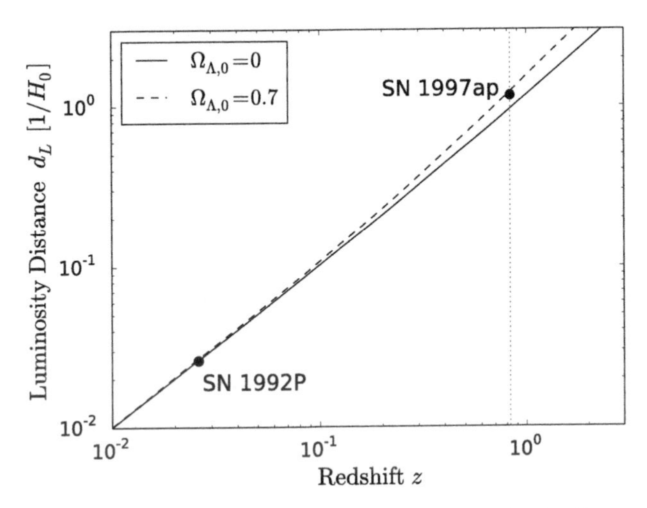

Figure 12.3 Luminosity distance d_L versus the redshift z plots for the flat matter-dominated Universe (solid line) and flat Universe with matter and dark energy corresponding to $\Omega_\Lambda = 0.7$ (dashed line). The two points are observed luminosity distances for the two Type-Ia supernovae: SN 1992P at $z = 0.026$ and SN 1997ap at $z = 0.83$.

Putting together in one model dark energy (denoted, as before, with Λ) and CDM leads to the ΛCDM cosmology, or the *standard model of cosmology*. The predictions of ΛCDM generally agree impressively with observations. However, there are several important discrepancies between predictions of the ΛCDM paradigm and observations of galaxies and their clustering, thereby creating a potential crisis for the ΛCDM picture, including:

- The *cuspy halo problem* (or the *core-cusp problem*): ΛCDM predicts that the central density slopes of galaxies are much steeper than what has been observed [36].
- The *missing satellites problem* (or the *dwarf galaxy problem*): the ΛCDM predicts a large number of small dwarf galaxies about one-thousandth the mass of the Milky Way. The number of these dwarf galaxies and their small halos that has been observed is orders-of-magnitude lower than expected from simulations [37–39].
- The *lithium problem* (or *lithium discrepancy*): ΛCDM predicts that the abundance of lithium should be about three times larger than what has been observed [40].
- The *Hubble tension* [41]: Within ΛCDM, the present values of the Hubble parameter H_0 obtained from the low-redshift Type-IA supernovae surveys [42] ($h \approx 0.73$) disagree to within 5σ confidence with the H_0 value from CMB observations ($h \approx 0.674$) [19].
- The *baryonic Tully–Fisher relation* (BTFR) finds that the baryonic mass in spiral galaxies is proportional to the value of flat rotation curves v_∞ to the power of roughly 3.5–4 [43]. If the dark matter dominates baryonic matter in spiral galaxies, as ΛCDM posits, it is unexpected that the kinematics of the entire galaxy would be so strongly correlated to the luminous component.
- The *radial acceleration relation* (RAR) shows that the observed and predicted accelerations in spiral galaxies are very tightly correlated, much more so than expected if dark matter dominated galaxy dynamics [44]. If dark matter were indeed the dominant matter component in galaxies, as assumed by the paradigmatic ΛCDM model, then no strong correlation should exist between the measured acceleration in spiral galaxies and that calculated using the nondominant visible-matter component in Newton's equations. Similarly as in BTFR, it seems as if the dynamics of spiral galaxies is determined more by its less abundant visible component than its supposedly dominant dark component. This presents a problem for the ΛCDM paradigm.

12.8.1 Detecting dark energy using luminosity distance vs. redshift

Let us illustrate how this direct evidence of the dark energy was obtained from the measurements of the luminosity distance for Type-Ia supernovae standard candles.

The luminosity distance d_L given by Eq. (10.17),

$$d_L = \frac{\chi}{a}, \tag{12.57}$$

where χ is the comoving distance defined in Eq. (10.6) as

$$\chi \equiv \int_{t(a)}^{t(0)} \frac{d\tilde{t}}{a(\tilde{t})} = \int_a^1 \frac{d\tilde{a}}{\tilde{a}^2 H(\tilde{a})} = \int_a^1 \frac{d\tilde{a}}{\tilde{a}^2 \left(\frac{\dot{\tilde{a}}}{\tilde{a}}\right)} = \int_a^1 \frac{d\tilde{a}}{\dot{\tilde{a}}\tilde{a}}, \tag{12.58}$$

after the change of variables $da/dt = aH$ and recalling $H \equiv \dot{a}/a$. Allowing for the non-zero cosmological constant Λ representing dark energy in addition to matter in a flat Universe ($\Omega_{\mathrm{T},0} = 1 = \Omega_{\mathrm{m},0} + \Omega_{\Lambda,0}$), we have from the first Friedmann equation (Eq. (9.21)):

$$\left(\frac{\dot{a}}{a}\right)^2 = H_0^2\left[(1-\Omega_{\Lambda,0})\frac{1}{a^3} + \Omega_{\Lambda,0}\right] \quad \rightarrow \quad \dot{a} = H_0\sqrt{(1-\Omega_{\Lambda,0})a^{-1} + \Omega_{\Lambda,0}a^2}. \tag{12.59}$$

After substituting into Eq. (12.58) above, we obtain

$$\chi(a) = \frac{1}{H_0}\int_a^1 \frac{d\tilde{a}}{\tilde{a}\sqrt{(1-\Omega_{\Lambda,0})\tilde{a}^{-1} + \Omega_{\Lambda,0}\tilde{a}^2}} = \frac{1}{H_0}\int_a^1 \frac{d\tilde{a}}{\sqrt{(1-\Omega_{\Lambda,0})\tilde{a} + \Omega_{\Lambda,0}\tilde{a}^4}}, \tag{12.60}$$

or, in terms of the redshift z, from the relation $a = 1/(1+z)$:

$$\chi(z) = \frac{1}{H_0}\int_0^z \frac{d\tilde{z}}{\sqrt{(1-\Omega_{\Lambda,0})(1+\tilde{z})^3 + \Omega_{\Lambda,0}}}. \tag{12.61}$$

The corresponding luminosity distance d_L is then given by

$$d_L(z) \equiv \frac{\chi(z)}{a} = \chi(z)(1+z) = \frac{1+z}{H_0}\int_0^z \frac{d\tilde{z}}{\sqrt{(1-\Omega_{\Lambda,0})(1+\tilde{z})^3 + \Omega_{\Lambda,0}}}, \tag{12.62}$$

which is what is used to obtain Fig. 12.3 and Fig. 12.4.

The *apparent magnitude* m measures a flux from an astronomical object as seen from Earth:

$$m = -\frac{5}{2}\log\left(\frac{f}{f_0}\right), \tag{12.63}$$

where f_0 is some reference flux.

The *absolute magnitude* M is defined as the apparent magnitude an astronomical object would have at a distance of 10 pc:

$$M = m - 5\log\left(\frac{d_L}{10\ \mathrm{pc}}\right). \tag{12.64}$$

When comparing apparent magnitudes m_1 and m_2 of the two objects of the same type – with the same absolute magnitude M (such as Type-Ia supernova) – the above equation is equivalent to

$$m_1 - m_2 = 5\log\left(\frac{d_L(m_1)}{10\ \mathrm{pc}}\right) - 5\log\left(\frac{d_L(m_2)}{10\ \mathrm{pc}}\right). \tag{12.65}$$

This is because of the way the magnitudes are defined: the difference of 5 magnitudes (mag) is equivalent to the brightness (flux) ratio of 100:

$$\frac{f_2}{f_1} = 100^{\frac{(m_1 - m_2)}{5}}, \tag{12.66}$$

where f_1 and f_2 are the fluxes of the two objects and m_1 and m_2 are their apparent magnitudes.

Example 1: The methodology of this kind of measurement can be well illustrated by considering the two supernovae from this sample: SN 1997ap at redshift $z_1 = 0.83$ with apparent magnitude $m_1 = 24.32$ and SN 1992P at redshift $z_2 = 0.026$ and apparent magnitude $m_2 = 16.08$. Since the absolute magnitudes of these are the same (because Type-Ia supernovae are "standard candles"), the difference in apparent magnitudes is entirely due to the difference in luminosity distance:

$$m_1 - m_2 = 5 \log\left(\frac{d_L(z_1)}{10 \text{ pc}}\right) - 5 \log\left(\frac{d_L(z_2)}{10 \text{ pc}}\right). \tag{12.67}$$

The second supernova (SN 1992P) is so close that its luminosity distance is unaffected by cosmology (see Fig. 12.3), and subscribes to the Hubble law valid for small redshifts z: $d_L = z/H_0$. The luminosity distance for SN 1992P is then given by $d_L(z_2) = z_2/H_0 = 0.026/H_0$. The only remaining unknown in Eq. (12.67) is fixed by observations to be

$$d_L(z = 0.83) = 1.16/H_0. \tag{12.68}$$

For a flat, matter-dominated Universe ($\Omega_T = \Omega_m = 1$), the luminosity distance at $z = 0.83$ is equal to $0.95/H_0$, while the Universe with $\Omega_{m,0} = 0.3$ and $\Omega_{\Lambda,0} = 0.7$ has the luminosity distance of $1.23/H_0$. Therefore the apparent magnitude of supernova SN 1997ap suggests that there is a sizable component of the dark energy.

Two research groups – one led by Saul Perlmutter [45] and the other led by Brian Schmidt [46] – measured the apparent magnitudes m for a large set of Type-Ia supernovae and established a systematic bias toward a Universe with a considerable contribution to the total energy density coming from dark energy, as illustrated in Fig. 12.4. The measurement of Type-Ia supernovae conducted by the two teams led to the constraints on the two free parameters: the current values of the relative content of matter ($\Omega_{m,0}$) and the dark energy ($\Omega_{\Lambda,0}$), which is only one of the possible ways to model it (the relative content of radiation $\Omega_{r,0}$ has been negligible compared to the other two components since the Universe was about 50,000 years old – well before recombination – and therefore does not affect the computation of the luminosity distance; see Appendix G). The leaders of the two groups, as well as Adam Riess, received the 2011 Nobel Prize in physics for their troubles [47].

A plot of luminosity distance containing data from several groups that have since measured the Type-Ia supernovae data is shown in Fig. 12.4.

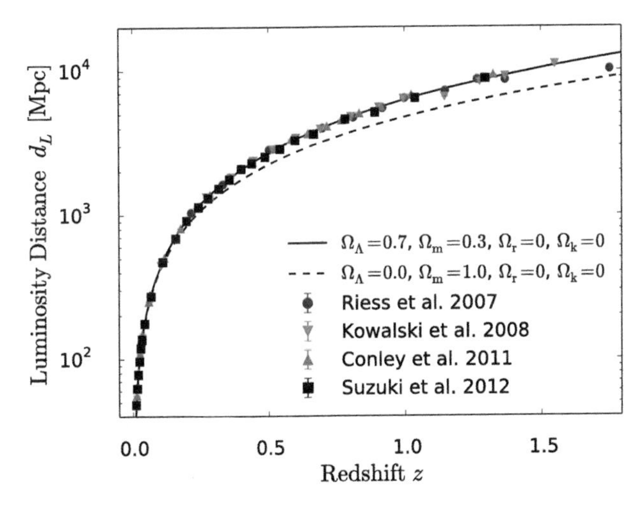

Figure 12.4 Luminosity distance d_L versus the redshift z plots for the flat matter-dominated Universe (solid line) and flat Universe with matter and dark energy corresponding to $\Omega_{\Lambda,0} = 0.7$ (dashed line). $h = 0.73$ is used. The data are from several sources [48–51].

12.8.2 Investigating other forms of dark energy

In order to investigate other forms of dark energy, we allow for dark-energy density to be time dependent (and not due to the cosmological constant Λ). The equation of state $P = P(\rho)$ for dark energy must obey Friedmann's second equation (Eq. (7.34c)):

$$\frac{d\rho}{dt} + 3\left(\rho + \frac{P}{c^2}\right)\frac{\dot{a}}{a} = 0. \tag{12.69}$$

For time-independent dark energy, *i.e.*, due to cosmological constant Λ, the equation of state is

$$P = -\rho c^2. \tag{12.70}$$

Earlier, we introduced a parameter w in the equation of state:

$$w \equiv \frac{P}{\rho c^2}, \tag{12.71}$$

where $w = 0$ for matter, $w = 1/3$ for radiation and $w = -1$ for dark energy due to the cosmological constant (see Table 11.1). The two studies of supernovae also computed the likelihood regions in the (Ω_{M}, w) space in the case of a flat Universe. The cosmological constant ($w = -1$) is allowed, but is not the only possibility.

To compute how the time-dependent dark-energy density, as denoted by $w = w(t)$, or equivalently $w = w(a)$, evolves with the expanding Universe, we can solve Eq. (12.69) with $w = w(a)$:

$$\frac{d\rho}{dt} = -3\left(\rho + w(a)\rho\right)\frac{\dot{a}}{a} = -3\rho\left(1 + w(a)\right)\frac{da}{adt} \quad \rightarrow \quad \int_0^a \frac{d\rho}{\rho} = -3\int_0^a \left(1 + w(a')\right)\frac{da'}{a'}$$
$$\rightarrow \quad \rho \propto \exp\left\{-3\int_0^a \left(1 + w(a')\right)\frac{da'}{a'}\right\}. \tag{12.72}$$

If $w = const.$, then

$$\rho \propto \exp\left\{-3\left(1 + w\right)\int_0^a \frac{da'}{a'}\right\} = \exp\left\{-3\left(1 + w\right)\ln a\right\} = \exp\left\{\ln\left(a^{-3(1+w)}\right)\right\} = a^{-3(1+w)}, \tag{12.73}$$

which matches $\rho \propto a^{-3}$ for $w = 0$ (matter: dust approximation $P = 0$), $\rho \propto a^{-4}$ for $w = 1/3$ (radiation: perfect-fluid approximation $P = \rho/3$), and $\rho \propto const.$ for $w = -1$ (cosmological constant Λ: $P = -\rho$).

There are several "popular" values of w for the dark energy:

- $w < -1/3$: *quintessence*;
- $w = -1$: *cosmological constant* Λ;
- $w < -1$: *phantom energy*.

12.8.3 Alternative to dark energy

One approach toward explaining what we perceive as dark energy is to revisit the underlying assumptions of our cosmological model and the resulting equations, most notably the assumption of "homogeneity" of the Universe. The Universe only appears homogeneous on the largest scales, while it has a complicated "Swiss cheese" structure whose expansion *differs* from the expansion of the homogeneous model. After revisiting Einstein's equations, one finds that the inhomogeneity generates a term analogous to the vacuum-energy term. It is still very much an open issue whether this term is of sufficient magnitude to cause the Universe to evolve in the manner we observe.

Exercises

1. Derive Eq. (12.14).
2. Derive Eq. (12.16).
3. Assuming blackbody radiation, convert $T_{\gamma,0} = 2.73$ K into
   a. eV;

b. $n_{\gamma,0}$ in photons cm^{-3};
c. $\rho_{\gamma,0}$ in photons eV cm^{-3};
d. $\rho_{\gamma,0}$ in photons g cm^{-3};
e. $\Omega_{\gamma,0}$.
Use $H_0 = 100h$ km s^{-1} Mpc^{-1} and $h = 0.73$.

CHAPTER 13

Brief history of the early Universe

The Universe is full of magical things, patiently waiting for our wits to grow sharper.

Eden Phillpots

The Big Picture: In this chapter we outline the standard model of the Universe in its early stages immediately following the hot Big Bang. These earliest epochs in the evolution of the Universe are still inadequately understood. As we move away from the Big Bang, our understanding of the physical epochs of the Universe improves considerably.

13.1. Keeping track of the Universe's history

The different times in the history of the Universe can be tracked by any of the several quantities that change monotonically throughout: the age of the Universe t, the scale factor a, the cosmological redshift (as we observe it today) z, and the temperature of the CMB radiation T_γ (currently measured at ≈ 2.73 K). From Eq. (12.21),

$$u_\gamma = a_r T_\gamma^4, \tag{13.1}$$

and the result derived from Friedmann's second equation that the radiation scales as

$$u_\gamma \propto a^{-4}, \tag{13.2}$$

we obtain that

$$T_\gamma(a) \propto a^{-1}, \tag{13.3}$$

which, combined with the current measurement ($a = 1$) of the temperature of the CMB radiation,

$$T_{\gamma,0} \approx 2.73 \text{ K}, \tag{13.4}$$

yields

$$T_\gamma(a) = T_{\gamma,0} a^{-1} \approx 2.73 a^{-1} \text{ K}. \tag{13.5}$$

To relate this to the age of the Universe t, one can explicitly solve integrals for $a(t)$ and substitute in Eq. (13.5). The relationships between any two quantities t, z, a, and T_γ are given in Table 13.1.

It is beneficial to relate directly – albeit crudely – the temperature T and the age of the Universe t. With the relationship between the temperature T and the scale of the

Relativity and Cosmology
https://doi.org/10.1016/B978-0-44-323542-9.00021-3

Universe a given in Eq. (13.5), all we need for $T(t)$ is $a(t)$, as $T(t) = T(a(t))$. Earlier, we have found $a(t)$ analytically by considering only the dominant terms for a matter-dominated or a radiation-dominated Universe. Relating the scale factor a and the age of the Universe t in a more general case when the Universe contains matter, radiation, and the cosmological constant representing dark energy requires solving the integral given in Table 13.1 for $t(a)$ and inverting it. This can only be done numerically. The analytic approximations for $a(t)$ we derived earlier are:

1. Flat, radiation-dominated Universe (Eq. (8.15)): $a(t) = (2H_0)^{1/2}\,t^{1/2} = \sqrt{2}\,(H_0 t)^{1/2}$,
2. Flat, matter-dominated Universe (Eq. (8.19)): $a(t) = \left(\frac{3H_0}{2}\right)^{2/3} t^{2/3} = (3/2)^{2/3}\,(H_0 t)^{2/3}$,

where

$$H_0 = 100\ h\ \text{km s}^{-1}\ \text{Mpc}^{-1} = 100h\left(\frac{1000\ \text{m}}{1\ \text{km}}\right)\left(\frac{1\ \text{Mpc}}{3.0856\times 10^{22}\ \text{m}}\right)\ \text{km s}^{-1}\ \text{Mpc}^{-1}$$
$$H_0 \approx 3.24\ h\times 10^{-18}\ \text{s}^{-1} \approx 2.37\times 10^{-18}\ \text{s}^{-1}, \tag{13.6}$$

with $h \approx 0.73$. Therefore for the two approximations, we have:

1. Flat, radiation-dominated Universe (Eq. (8.15)): $a(t) = 2.2\times 10^{-10}\ t^{1/2}$,
2. Flat, matter-dominated Universe: $a(t) = 2.3\times 10^{-12}\ t^{2/3}$,

where t is measured in seconds. When these are combined with Eq. (13.5), we obtain:

1. Flat, radiation-dominated Universe: $T_\gamma(t) \approx 10^{10}\ t^{-1/2}$ K,
2. Flat, matter-dominated Universe: $T_\gamma(t) \approx 10^{12}\ t^{-2/3}$ K,

where, again, t is measured in seconds.

Table 13.1 Relationship between the scale of the Universe (a), age of the Universe (t), redshift as observed from here today (z), and the temperature of the CMB radiation T_γ.

Quantity		Dependence on scale a	Dependence on redshift z
Age	t	$t(a) = \frac{1}{H_0}\int_0^a \frac{\tilde{a}d\tilde{a}}{\sqrt{\Omega_{r,0}+\Omega_{m,0}\tilde{a}+\Omega_\Lambda\tilde{a}^4}}$	$t(z) = \frac{1}{H_0}\int_z^\infty \frac{d\tilde{z}}{\sqrt{\Omega_{r,0}(1+\tilde{z})^6+\Omega_{m,0}(1+\tilde{z})^5+\Omega_\Lambda(1+\tilde{z})^2}}$
Redshift	z	$z(a) = 1/a - 1$	–
Scale	a	–	$a(z) = 1/(1+z)$
Temp.	T_γ	$T_\gamma(a) = 2.7a^{-1}$ K	$T_\gamma(z) = 2.7(1+z)$ K

13.2. Early Universe at a glance: major benchmarks

As the Universe evolves from the Big Bang and its temperature decreases, a recurrent theme emerges: when the Universe's temperature drops below the mass of the particles in thermal equilibrium, particle–anti-particle pair creation stops and the particle's number density precipitously drops.

Before we delve into some detail on different epochs in the early Universe, it is instructive to outline broadly the major benchmarks between the Big Bang ($t = 0$) and recombination ($t \approx 400,000$ years).

- $T > 10^{32}$ K: Quantum physics predicts the breakdown of spacetime.
- $T > 10^{12}$ K: The Universe is dominated by physical processes that have not been tested in humankind's laboratories, but one can at least assume (it is believed) that the Universe was hot and dense, and that the dynamics is dominated by massless, or nearly massless, particles, *i.e.*, $kT \gg mc^2$.
- $T \sim 10^{11} - 10^{12}$ K: The Universe is filled with photons (γ), neutrinos, and anti-neutrinos (ν and $\bar{\nu}$), muons and anti-muons ($\mu^{\pm}$), electrons and positrons ($e^{\pm}$), and nucleons (p and n), all maintained in an equilibrium because of the effects of the electroweak and strong interactions.
- $T \sim 1 - 3 \times 10^{10}$ K: Neutrinos decouple from everything else because the weak interaction is too weak to maintain equilibrium; the proton:neutron ratio falls out of equilibrium because of beta decay.
- $T \sim 3 \times 10^{9}$ K: Electron–positron pair formation ends and the remaining pairs simply annihilate via $e^+ + e^- \rightarrow 2\gamma$ or $e^+ + e^- \rightarrow 3\gamma$.
- $T \sim 10^{9}$ K: Nucleosynthesis begins, resulting in the formation of light elements like ^{4}He, D, ^{4}He, Li, and Be.
- $T \sim 3 \times 10^{8}$ K: Nucleosynthesis ends.
- $T > 10^{3} - 10^{5}$ K: The energy density of the Universe continues to be dominated by neutrinos and photons, which cool with $T_\gamma \propto a^{-1}$; the nucleons remain coupled to the photons because of electromagnetism and thus share the same temperature.
- $T \sim 10^{3} - 10^{5}$ K: The energy density of the photons and neutrinos becomes smaller than the rest-mass energy of nucleons, and the Universe passes from the *radiation-dominated* epoch to the *matter-dominated* epoch.
- $T \sim 4000$ K: Electrons and nuclei combine to form neutral atoms (*recombination*). The atoms and photons are effectively decoupled and subsequently evolve independently. The primordial photons constitute the CMB whose temperature evolves as $T_\gamma \propto a^{-1}$.

13.3. *Very* early Universe: the first three minutes

The correct fundamental physics of the very early Universe is unknown, although a variety of options have been proposed, including grand unified theories (GUTs), inflation, string theory, supersymmetry, phase transition, quantum gravity, and theories of everything (TOEs). In this section, we describe the history of the *very* early Universe, and summarize it in Table 13.2.

13.3.1 The Big Bang: $t = 0$ s

Extrapolation of the expansion of the Universe backward in time using general relativity yields an infinite density and an infinite temperature at a finite time in the past. This singularity signals the breakdown of general relativity. How closely we can extrapolate toward the singularity is debated – certainly not earlier than the Planck epoch (see next section). The early hot, dense phase is itself referred to as *the Big Bang* and is considered as the "birth" of our Universe – The Beginning.

The discussion about the nature, cause, and origin of the Big Bang itself is at best very difficult, if not untestable, and as such quickly enters the realms of metaphysics or theology.

13.3.2 The Planck epoch: $0 < t \lesssim 10^{-43}$ s

The Planck epoch is the earliest period of time in the history of the Universe, spanning the brief time immediately following the Big Bang during which the quantum effects of gravity were significant.

In order to compute the time-scale over which quantum effects dominate (barring the existence of branes that would circumvent them), we use dimensional analysis:

Effects	Constant	Value	Units
Relativity	c	3.00×10^{10}	$\frac{\text{cm}}{\text{s}}$
Quantum mechanics	$\hbar$	1.06×10^{-27}	$\frac{\text{g cm}^2}{\text{s}}$
Gravitation	G	6.67×10^{-8}	$\frac{\text{cm}^3}{\text{g s}^2}$

We need to find the way to combine the constants above to obtain the relevant time-scale:

$$c^A \hbar^B G^D = \text{s},$$

$$\rightarrow \quad \left(\frac{\text{cm}}{\text{s}}\right)^A \left(\frac{\text{g cm}^2}{\text{s}}\right)^B \left(\frac{\text{cm}^3}{\text{g s}^2}\right)^D = \text{s},$$

$$\begin{array}{rcccl}
[\text{cm}]: & A & +2B & +3D & = 0 \\
[\text{g}]: & & +B & -D & = 0 \\
[\text{s}]: & -A & -B & -2D & = 1 \\
\hline
\text{Solution} & A = -\frac{5}{2} & B = \frac{1}{2} & D = \frac{1}{2} & \rightarrow \quad t_P = c^{-\frac{5}{2}} \hbar^{\frac{1}{2}} G^{\frac{1}{2}}.
\end{array}$$

Therefore the time-scale for quantum gravity, the *Planck time* t_P, is

$$t_P \equiv \sqrt{\frac{\hbar G}{c^5}} = 5.39 \times 10^{-44} \text{ s.} \tag{13.7}$$

In a similar fashion, we can find the *Planck length*:

$$l_P \equiv \sqrt{\frac{\hbar G}{c^3}} = ct_P = 1.62 \times 10^{-33}\ \text{cm}, \tag{13.8}$$

and the *Planck energy*:

$$E_P \equiv \sqrt{\frac{\hbar c^5}{G}} = 1.22 \times 10^{19}\ \text{GeV}. \tag{13.9}$$

The Planck energy can be directly translated into the temperature T via $E = kT$, yielding the *Planck temperature*:

$$T_P = 1.42 \times 10^{32}\ \text{K}. \tag{13.10}$$

If the supersymmetry hypothesis is correct, then during this time the four fundamental forces – electromagnetism, weak force, strong force, and gravity – all have the same strength, so they are possibly unified into one fundamental force. Our understanding of this early epoch is still quite tenuous, patiently awaiting a happy marriage of quantum mechanics and relativistic gravity.

13.3.3 Grand unification epoch: $10^{-43}\ \text{s} \lesssim t \lesssim 10^{-36}\ \text{s}$

The Grand Unification Epoch immediately followed the Planck Epoch. The temperature of the Universe was on the order of those that characterize Grand Unification Theories (GUTs), in which grand unification energies are thought to be $\sim 10^{15}$ GeV, corresponding to temperatures exceeding 10^{28} K. During the Grand Unification Epoch, three of the four fundamental forces – electromagnetism, the strong force, and the weak force – were possibly unified as the electronuclear force. Gravity had detached from the electronuclear force at the end of the Planck era.

The Grand Unification Epoch finished when the strong force separated from the other fundamental forces, at about 10^{-36} s.

13.3.4 Inflationary epoch: $10^{-36}\ \text{s} \lesssim t \lesssim 10^{-32}\ \text{s}$

In Section 9.5, we discussed different scenarios in the early Universe that allowed for all matter to be causally connected, and therefore initially in equilibrium. Here, again, we put focus on the cosmological inflation.

If inflation theory is correct, the Inflationary Epoch immediately followed the Grand Unification Epoch. During the Inflationary Epoch, the Universe experienced an exponential expansion, which increased the radius of the Universe by a factor of at least 10^{26}, corresponding to the growth in volume of at least 10^{78}. At this time, the strong force started to separate from the electroweak interaction.

The expansion is thought to have been triggered by the phase transition that marked the end of the preceding Grand Unification Epoch at approximately 10^{-36} s after the Big Bang. One of the theoretical products of this phase transition was a scalar field called the inflaton field. As this field settled into its lowest-energy state throughout the Universe, it generated a repulsive force that led to a rapid expansion of the fabric of spacetime. This expansion explains various properties of the current Universe that are difficult to account for without the Inflationary Epoch (for example, the flat Universe or the horizon problem).

The rapid expansion of spacetime meant that elementary particles remaining from the Grand Unification Epoch were now distributed very thinly across the Universe. However, the huge potential energy of the inflation field was released at the end of the Inflationary Epoch, repopulating the Universe with a dense, hot mixture of quarks, anti-quarks, gluons, leptons, and photons as it entered the Electroweak Epoch.

13.3.5 Electroweak epoch: $10^{-32}\,\mathrm{s} \lesssim t \lesssim 10^{-12}\,\mathrm{s}$

During the Electroweak Epoch, which followed the Inflationary Epoch, the temperature of the Universe was high enough to merge electromagnetism and the weak interaction into a single electroweak interaction (≈ 100 GeV $\approx 10^{15}$ K). At approximately 10^{-32} s after the Big Bang the potential energy of the inflation field that had driven the inflation of the Universe during the Inflationary Epoch was released, filling the Universe with a dense, hot quark–gluon plasma (*reheating*). Particle interactions in this phase were energetic enough to create large numbers of other particles, including W, Z, and Higgs bosons. As the Universe expanded and cooled, interactions became less energetic, and when the Universe was about 10^{-12} s old, W, Z, and Higgs bosons ceased to be created. The remaining W, Z, and Higgs bosons decayed quickly. The weak interaction became a short-range force in the following Quark Epoch.

After the Inflationary Epoch, the physics of the Electroweak Epoch is less speculative and better understood than for previous periods of the early Universe. The existence of W and Z bosons has been demonstrated, and other predictions of electroweak theory have been experimentally verified.

13.3.6 Quark epoch: $10^{-12}\,\mathrm{s} \lesssim t \lesssim 10^{-6}\,\mathrm{s}$

During the Quark Epoch, which started about 10^{-12} s after the Big Bang, the fundamental interactions of gravitation, electromagnetism, the strong interaction, and the weak interaction had all taken on their present, separate forms. The temperature of the Universe was still too high to allow quarks to bind together to form hadrons, resulting in a dense, hot quark–gluon plasma composed of quarks, gluons, leptons, and photons. Quarks that collided had too much energy to permit formation of mesons or baryons. The Quark Epoch ended at about 10^{-6} s after the Big Bang, when the average energy of particle interactions dropped below the binding energy of hadrons. This allowed for the hadrons to start forming, ushering in the Hadron Epoch.

13.3.7 Hadron epoch: $10^{-6}\ \mathrm{s} \lesssim t \lesssim 1\ \mathrm{s}$

The Hadron Epoch was the period in the evolution of the early Universe during which the mass of the Universe was dominated by hadrons. It started approximately 10^{-6} s after the Big Bang, when the Universe had cooled enough to allow the quarks to bind together into hadrons. Initially, the Universe was hot enough to allow the creation of hadron–anti-hadron pairs, thereby keeping matter and anti-matter in thermal equilibrium. However, as the Universe continued to cool, the production of hadron–anti-hadron pairs stopped. The vast majority of the hadrons and anti-hadrons annihilated each other, leaving a small residue of hadrons. The elimination of anti-hadrons was completed by 1 second after the Big Bang, when the following Lepton Epoch began.

13.3.8 Lepton epoch: $1\ \mathrm{s} \lesssim t \lesssim 3\ \mathrm{min}$

From the Planck time t_P of quantum gravity up to the Lepton Epoch, the physics of the Universe is dominated by *very high* temperatures ($> 10^{12}$ K) and therefore by high-energy particle physics.

- **Muon–anti-muon annihilation**:
  At sufficiently high temperatures, there is balance of *pair production* ($\rightarrow$) and *muon–anti-muon annihilation* ($\leftarrow$):

$$\begin{aligned} &\gamma + \gamma \rightleftarrows \mu^+ + \mu^-, \\ \rightarrow \quad &\text{photon energy} \rightleftarrows \text{muon mass}. \end{aligned} \tag{13.11}$$

  The balance can persist only as long as $kT \approx 2m_\mu c^2$, where m_μ is the muon mass, so that

$$T \geq \frac{2m_\mu c^2}{k} = \frac{2(200m_e)c^2}{k} = \frac{2(2009.1 \times 10^{-28})(3 \times 10^{10})^2}{1.38 \times 10^{-16}} = 2 \times 10^{12}\ \mathrm{K}. \tag{13.12}$$

  Below this temperature, no pair production can take place, and muons and anti-muons annihilate at around $T \approx 10^{12}$ K.

- **Electron–positron annihilation**:
  The balance between electron–positron pair production and electron–positron annihilation is maintained until around

$$T \geq \frac{2m_e c^2}{k} \approx 10^{10}\ \mathrm{K}, \tag{13.13}$$

  below which electrons and positrons annihilate.

Table 13.2 Very early history of the Big Bang Universe, up to the Big Bang Nucleosynthesis. Recall that theories other than cosmological inflation (as discussed in Section 9.5) can put matter in causal contact very early in the history of the Universe. Temperature estimates are based on the matter-dominated Universe approximation: $T(t) \approx 10^{12} t^{-2/3}$ K (with 1 K $= 11\,605$ eV $\approx 10^4$ eV).

Epoch	Temperature	Characteristics
Big Bang 0 s	∞ K ∞ eV	singularity (vacuum fluctuation?)
Planck $0\text{ s} < t \leq 10^{-43}$ s	$> 10^{40}$ K $> 10^{36}$ eV	quantum gravity
Grand Unification $10^{-43}\text{ s} \leq t \leq 10^{-36}$ s	$10^{36} - 10^{40}$ K $10^{26} - 10^{32}$ eV	gravity freezes out the "grand unified force" (GUT)
Inflationary (?) $10^{-36}\text{ s} \leq t \leq 10^{-32}$ s	$10^{33} - 10^{36}$ K $10^{29} - 10^{32}$ eV	inflation begins strong force freezes out
Electroweak $10^{-32}\text{ s} \leq t \leq 10^{-12}$ s	$10^{20} - 10^{33}$ K $10^{16} - 10^{29}$ eV	weak force freezes out 4 distinct forces (EM dominates) **baryogenesis:** baryons and anti-baryons annihilate
Quark $10^{-12}\text{ s} \leq t \leq 10^{-6}$ s	$10^{16} - 10^{20}$ K $10^{12} - 10^{16}$ eV	Universe contains hot quark–gluon plasma: quarks, gluons, and leptons
Hadron $10^{-6}\text{ s} \leq t \leq 1$ s	$10^{12} - 10^{16}$ K $10^{8} - 10^{12}$ eV	quarks and gluons bind into hadrons
Lepton $1\text{ s} \leq t \leq 3$ min	$10^{10} - 10^{12}$ K $10^{6} - 10^{8}$ eV	Universe contains photons (γ), muons ($\mu^{\pm}$), electrons/positrons ($e^{\pm}$), and neutrinos (ν, $\bar{\nu}$); nucleons n and p in equal numbers
1 s	$\leq 10^{12}$ K $\leq 10^{8}$ eV	μ^+ and μ^- annihilate; ν and $\bar{\nu}$ decouple; $e^{\pm}$, γ and nucleons remain. Reactions: $e^+ + n \rightleftharpoons p + \nu_e$ $e^- + p \rightleftharpoons n + \nu_e$ $n \to p + e^- + \bar{\nu}_e$
100 s	10^{10} K, 10^{6} eV	e^+ and e^- annihilate

- **Decoupling of electron neutrinos**:
  Assuming a matter-dominated Universe, we crudely estimate electron number density:

$$
\begin{aligned}
\rho &= \rho_0 \left(\frac{a_0}{a}\right)^3 = \frac{\rho_0}{\left(2.3 \times 10^{-12}\, t^{2/3}\, \text{s}^{-2/3}\right)^3} \approx \frac{10^{-29}\ \text{g cm}^{-3}}{10^{-35}\, t^2\, \text{s}^{-2}} = 10^6\, t^{-2}\, \text{s}^2\ \text{g cm}^{-3}, \\
n_e &= \frac{\rho}{m_e} \approx \frac{10^6\, t^{-2}\, \text{s}^2\ \text{g cm}^{-3}}{9.1 \times 10^{-28}\ \text{g}} \approx \frac{10^{33}\ \text{s}^2\ \text{g cm}^{-3}}{t^2}.
\end{aligned}
\tag{13.14}
$$

The neutrino scattering cross section at $T = 10^{12}$ K is $\sigma_\nu \approx 10^{-44}$ cm^2, so the time between scatterings is

$$t_\nu \approx \frac{1}{n_e \sigma_\nu c}. \tag{13.15}$$

Scatterings will become "scarce" when

$$t_\nu \approx \frac{1}{\frac{10^{33}\ \mathrm{s}^2\ \mathrm{g\ cm}^{-3}}{t_\nu^2}\sigma_\nu c} \quad \rightarrow \quad t_\nu = \left(10^{33}\ \mathrm{s}^2\ \mathrm{g\ cm}^{-3}\right)\sigma_\nu c \approx 0.3\ \mathrm{s}. \tag{13.16}$$

Therefore electron neutrinos decouple from the other matter in the Universe at about $t \approx 1$ s.

13.4. Early Universe: the first 400,000 years

13.4.1 The earliest Universe: $T \gtrsim 10^{12}$ K

The one thing about which essentially all physicists agree is that, at these very high temperatures, the dominant energy source comprised massless and nearly massless particles, where 'nearly massless' means that their rest-mass energy is much smaller than their thermal energy, *i.e.*, $mc^2 \ll kT$. Because the temperature is so high, one anticipates the spontaneous generation of particle–anti-particle pairs, so that, for example, at $T \sim 10^{12}$ K, the Universe should contain huge numbers of pairs like $e^\pm$ and $\mu^\pm$, as well as ν and $\bar{\nu}$. There should be a slight excess in electrons over positrons, reflecting the fact that, today, the Universe is dominated by matter rather than anti-matter. The number of protons should equal the net excess of electrons over positrons, given the expectation that, consistent with experiments, the Universe is electrically neutral. There should be comparable numbers of protons and neutrons, since they have approximately the same mass. At temperatures as low as 10^{12} K, there should *not* be large numbers of anti-protons and anti-neutrons, since their rest-mass energy $mc^2 \gg kT$.

As recently as 1998, there was no compelling reason experimentally to believe that neutrinos are not exactly massless. However, there is now evidence that neutrinos indeed have non-zero rest masses. Nevertheless, these masses appear to be so small that, for all practical purposes, they behave as massless particles.

At these very high temperatures, all the different particle species should be in equilibrium with one another as a result of the electromagnetic and weak interactions. $e^\pm$ and $\mu^\pm$ pairs exist in a thermal balance because of the pair production and annihilation interactions

$$e^+ + e^- \leftrightarrow 2\gamma \quad \text{and} \quad \mu^+ + \mu^- \leftrightarrow 2\gamma, \tag{13.17}$$

and

$$e^+ + e^- \leftrightarrow 3\gamma \quad \text{and} \quad \mu^+ + \mu^- \leftrightarrow 3\gamma. \tag{13.18}$$

The two-photon processes involve direct annihilation. The three-photon processes involve bound entities like positronium (an 'atom' comprised of one electron and a positron bound via electricity). The relative number of electrons, protons, and neutrons is fixed by the interaction

$$n \leftrightarrow p + e^- + \bar{\nu}_e, \tag{13.19}$$

which involves the weak interaction.

The relative abundance of protons and neutrons is fixed by the *Saha equation*, which essentially introduces a Boltzmann factor reflecting the different energies of the proton and neutron:

$$\frac{n_n}{n_p} = \exp\left(-\frac{\Delta mc^2}{kT}\right), \tag{13.20}$$

where $\Delta m = m_n - m_p$ is the difference between the neutron m_n and proton mass m_p.

How dense was the Universe back then? Let us assume that the number of photons has remained roughly constant, $n_\gamma \propto a^{-3}$. Given, however, that $T_\gamma \propto a^{-1}$, it then follows that $n_\gamma \propto T_\gamma^3$, *i.e.*,

$$n_\gamma(t) = n_{\gamma,0}\left(\frac{T}{T_0}\right)^3. \tag{13.21}$$

Using $T = 10^{11}$ K, $T_0 = 2.7$ K and $n_{\gamma,0} = 420/\text{cm}^3$ (see Eq. (14.10)) yields a typical distance between photons (mean-free path),

$$l(T) \sim n_\gamma^{-1/3} \sim n_{\gamma,0}^{-1/3}\left(\frac{T_0}{T}\right) \text{ cm} \sim 4 \times 10^{-10}\left(\frac{10^{11}\text{ K}}{T}\right) \text{ cm} \sim 4000\left(\frac{10^{11}\text{ K}}{T}\right) \text{ fermi}, \tag{13.22}$$

where 1 fermi $= 10^{-13}$ cm. The distance between photons is much larger than the size of a nucleus, and since protons and neutrons are less abundant than photons, the distance between these particles is even greater. It follows that one need not worry about quark physics, *i.e.*, the details associated with the strong interaction. For all intents and purposes, neutrons and protons are individual particles rather than combinations of more fundamental building blocks.

At slightly lower temperatures, photon energies are too low to make appreciable numbers of $\mu^\pm$ pairs. The existing pairs decay into photons, and the photons re-equilibrate themselves with the remaining species of matter.

13.4.2 Neutrino decoupling: $T \approx 3 \times 10^{10}$ K

As the temperature of the Universe drops, the number density of various particles also decreases, so they interact less frequently. What this means is that the effects of the fundamental interactions of nature are becoming weaker.

At a temperature $T \sim 3 \times 10^{10}$ K, the weak interaction becomes sufficiently weak that neutrinos and anti-neutrinos effectively decouple from the rest of the matter and evolve independently. At this time, the energy density of the Universe is associated primarily with photons, neutrinos, and electron–positron pairs, which yield a total

$$u \approx u_\gamma + u_{\nu_e} + u_{\bar{\nu}_e} + u_{\nu_\mu} + u_{\bar{\nu}_\mu} + u_{\nu_\tau} + u_{\bar{\nu}_\tau} + u_{e^-} + u_{e^+} \\ = a_r T_\gamma^4 + 6 \times \frac{7}{16} a_r T_\nu^4 + 2 \times \frac{7}{8} a_r T_e^4, \tag{13.23}$$

where a_r is the radiation constant defined in Eq. (12.22). Here, the electrons and positrons are characterized by a factor 7/8 since they have a g-factor $g = 2$ (reflecting 2 spin states), unlike neutrinos, which have $g = 1$. It follows that, if the different particle species all have the same temperature T – as they do at this time because they are all in equilibrium – the energy density of the Universe is

$$u \approx \left(1 + 6 \times \frac{7}{16} + 2 \times \frac{7}{8}\right) a_r T^4 = \frac{43}{8} a_r T^4 = \frac{43}{8} \rho_\gamma. \tag{13.24}$$

If particle physicists were to discover that there are N_ν neutrino flavors, with $N_\nu \neq 3$, one would have instead that

$$u = \left(\frac{43}{8} + \frac{7}{8}(N_\nu - 3)\right) \rho_\gamma. \tag{13.25}$$

13.4.3 Neutron:proton ratio freezes out: $T \approx 10^{10}$ K

At temperature below $T \sim 10^{10}$ K, the weak interaction

$$n \rightleftarrows p + e^- + \bar{\nu} \tag{13.26}$$

can no longer go in both directions and therefore maintain an equilibrium between neutrons and protons. This means that, rather than decreasing with temperature in a fashion described by the Saha equation, the neutron to proton ratio becomes 'frozen out' at a value

$$\left.\frac{n_n}{n_p}\right|_* = \exp\left(-\frac{\Delta mc^2}{kT_*}\right), \tag{13.27}$$

where $\Delta mc^2 \approx 1.3$ MeV. Numerical simulations indicate that this 'freeze' occurs at $kT_* \approx 0.8$ MeV $\approx 0.9 \times 10^{10}$ K, which implies that

$$\left.\frac{n_n}{n_p}\right|_* \approx \frac{1}{5}. \tag{13.28}$$

After this point, the only thing that alters this ratio is the fact that free neutrons decay. A neutron tucked safely into an atomic nucleus is radioactively stable, but a free neutron not bound into a nucleus will decay with a half-life of approximately 10 min. What this means is that *if the epoch of nucleosynthesis did not occur until a time much longer than 10 min after the Big Bang, the Universe would contain essentially no neutrons and no complex nuclei could have been formed.*

13.4.4 Electron–positron annihilation: $T \approx 3 \times 10^9$ K

At a temperature $T \approx 3 \times 10^9$ K, the temperature is so low that photons cannot form electron–positron pairs. The net result is that the existing pairs annihilate into two or three photons, thereby pumping energy back into the electromagnetic radiation. What this means is that the photons will be reheated to a temperature $T_<$, which is slightly higher than the temperature $T_>$ that they would otherwise have had, thus ending up somewhat hotter than the neutrinos. This was obtained in Section 12.5, where we derived the increase in temperature by arguing conservation of entropy for the photons and electron–positron pairs.

13.4.5 The epoch of nucleosynthesis: $10^9 \text{ K} \gtrsim T \gtrsim 3 \times 10^8$ K

For $T \gtrsim 10^9$ K, there are so many high-energy photons that any two nucleons that have stuck together will immediately be broken apart as a result of collision with a photon. Hence, nucleosynthesis cannot begin.

For $T \lesssim 3 \times 10^8$ K, the nucleons are moving with sufficiently small kinetic energy that they cannot overcome Coulomb repulsion that exists between charges with the same sign. If, for example, two protons become very close to each other at $r \lesssim 1$ fermi, the strong interaction holds them together quite effectively. If, however, they are farther apart, the strong interaction becomes much weaker and is completely overpowered by the repulsive effect of the electromagnetic interaction. (In a Newtonian approximation, the electrostatic potential energy associated with a pair of protons $V_e \propto e^2/r$, whereas the strong interaction potential energy $V_s \propto (e^2/\alpha)\exp(-\kappa r)/r$, where $\alpha = 1/137$ is the fine structure constant and $\kappa^{-1} \approx 1$ fermi $= 10^{-13}$ cm.)

13.4.6 Transition from radiation-dominated Universe to matter-dominated Universe: $T \approx 10,000$ K

When did the Universe transition from the radiation to matter dominance? In other words, at what temperature T did $\rho_r \approx \rho_m$? Matter energy density is $u_m = \rho_m \approx m_p c^2 n_m$, but $n_m \propto a^{-3} \propto T_\gamma^3$, so that

$$u_m(T_\gamma) = u_m(T_0)\left(\frac{T_\gamma}{T_0}\right)^3 = u_{m,0}\left(\frac{T_\gamma}{T_0}\right)^3, \tag{13.29}$$

where $T_0 = 2.73$ K is the temperature of the photon CMB radiation. Similarly, we can express

$$u_r(T_\gamma) = u_r(T_0)\left(\frac{T_\gamma}{T_0}\right)^4 = u_{r,0}\left(\frac{T_\gamma}{T_0}\right)^4. \tag{13.30}$$

Setting $u_r = u_m$ to define the radiation/matter transition leads to

$$1 = \frac{u_r(T_\gamma)}{u_m(T_\gamma)} = \frac{u_{r,0}\left(\frac{T_\gamma}{T_0}\right)^4}{\rho_{m,0}\left(\frac{T_\gamma}{T_0}\right)^3} = \frac{u_{r,0}}{\rho_{m,0}}\left(\frac{T_\gamma}{T_0}\right) \tag{13.31}$$

and

$$T_\gamma = T_0\frac{u_{m,0}}{u_{r,0}}. \tag{13.32}$$

The energy density due to radiation u_r is composed of the energy density of the CMB photons u_γ and CMB neutrinos u_ν: $u_r = u_\gamma + u_\nu$. Using Eq. (13.24) without the contribution of electron–positron pairs and Eq. (12.37) to relate T_γ and T_ν as $T_\gamma = 1.4T_\nu$, leads to

$$u_r = u_\gamma + u_\nu = a_r T_\gamma^4 + 6\times\frac{7}{16}a_r T_\nu^4 = \left(1 + 6\times\frac{7}{16}\left(\frac{1}{1.4}\right)^4\right)a_r T_\gamma^4 \approx 1.68u_\gamma, \tag{13.33}$$

hence, $u_{r,0} = u_{\gamma,0} + u_{\nu,0} = 1.68u_{\gamma,0}$.
The temperature at the transition between radiation- and matter-dominance in Eq. (13.32) then becomes

$$T_\gamma = \frac{u_{m,0}}{1.68u_{\gamma,0}}T_0 = \frac{\Omega_{m,0}}{1.68\Omega_{\gamma,0}}T_0. \tag{13.34}$$

Using values $\Omega_{m,0} = 0.3$ and $\Omega_{\gamma,0} = 5\times10^{-5}$, we obtain $T_\gamma \approx 9600$ K.

At temperatures substantially above this value, it seems appropriate to assume an equation of state $P = u/3$ (ideal-gas approximation – radiation-dominated Universe). At temperatures substantially below this value, it is appropriate instead to assume $P = 0$ (dust approximation – matter-dominated Universe).

13.4.7 Recombination: $T \approx 4000$ K

The ground state of a hydrogen atom corresponds to a binding energy $E \approx 13.6$ eV. This means that any neutral hydrogen atom that forms can be completely disrupted by an interaction with a photon with energy $E > 13.6$ eV. Given, however, that an energy of 1 eV corresponds to a temperature $E = kT$ satisfying $T = 1.16 \times 10^4$ K, an energy $E = 13.6$ eV corresponds to a temperature $T = 1.58 \times 10^5$ K. One might thus anticipate that almost no atomic/neutral hydrogen would exist for $T \gg 10^5$ K, but that the abundance would begin to increase sharply once the temperature drops below $T \approx 10^5$ K. In fact, since a temperature only corresponds to the average energy of all the particles and it has a finite-width distribution, significantly lower temperatures are required before large amounts of neutral hydrogen can persist. As there are so many more photons than nucleons, even a tiny fraction of the photons having $E > 13.6$ eV corresponds to a comparatively high density of photons with energy large enough to ionize hydrogen. This means that one requires a temperature $T \ll 10^5$ K before neutral hydrogen can form.

To determine when neutral hydrogen begins to form with significant abundances requires a numerical computation that makes some assumption about the overall density of baryonic matter in the Universe. What that entails is writing $\rho_b = 3H_0^2\Omega_b/8\pi$ and exploring the effects of $h^2\Omega_b$. For $\Omega_b h^2 = 1$, one infers a fractional ionization of 99.8% at $T \approx 5000$ K, and a fractional ionization of 0.98% at $T = 3000$ K. The fractional ionization is 50% at $T \approx 4100$ K, which yields

$$
\begin{aligned}
T \propto a^{-1} \quad &\rightarrow \quad Ta = T_0 a_0 = const \quad \rightarrow \quad T = T_0\left(\frac{a_0}{a}\right) = T_0 a^{-1}, \\
T = T_0(1+z) \quad &\rightarrow \quad 1+z = \frac{T}{T_0} \approx \frac{4000}{2.7} \approx 1480 \approx 1500.
\end{aligned}
\tag{13.35}
$$

For $\Omega_b h^2 = 0.013$, the fractional ionization is 50% at $T \approx 3700$ K, which corresponds to a redshift satisfying $1 + z \approx 1370$. It is generally accepted that recombination ended at about $T \approx 3000$ K.

The 'size of the Universe' at the end of recombination t_R, as quantified by the scale factor a, satisfied

$$
\frac{a(t_R)}{a_0} = \frac{T_0}{T(t_R)} \approx \frac{2.73}{3000} \approx \frac{1}{1100}. \tag{13.36}
$$

It follows that the separation between two objects at the end of recombination was about 1100 times smaller than at the present time, and that the volume of a given piece of the Universe containing a fixed amount of stuff, was smaller by a factor of $\approx (1/1500)^3 \approx 7.5 \times 10^{-10}$.

Recombination was an event of crucial importance in the history of the Universe, since it was only after that epoch that matter could begin to evolve the large structures that are observed today: galaxies, groups of galaxies, galaxy clusters, and so forth via the so-called *Jeans instability* (see Appendix K).

The minimum scale on which matter can clump is the *Jeans length*,

$$\lambda_J \approx \sqrt{\frac{\pi c_s^2}{G\bar{\rho}}}, \tag{13.37}$$

where $\bar{\rho}$ is the mean mass density and c_s denotes a typical *speed of sound* (the speed at which density waves travel).

The speed of sound in the early Universe is

$$c_s = \sqrt{\frac{dP}{d\rho}} = \frac{c}{\sqrt{3}}, \tag{13.38}$$

(the derivation of this expression is left for the exercises in the next chapter) so that the Jeans length was very large:

$$\lambda_J \approx \frac{c}{\sqrt{G\bar{\rho}}}. \tag{13.39}$$

This length scale is comparable to the horizon length,

$$R_h \approx ct_h \approx \frac{c}{\sqrt{G\bar{\rho}}}, \tag{13.40}$$

the total distance light could have traveled in the entire history of the Universe t_h neglecting all interactions with matter. After recombination,

$$c_s \approx \sqrt{\frac{kT}{m}}, \tag{13.41}$$

which corresponds to a much shorter Jeans length.

Exercises

1. Derive the Planck energy given in Eq. (13.9).

CHAPTER 14

Cosmic microwave background radiation

Innocent light-minded men, who think that astronomy can be learnt by looking at the stars without knowledge of mathematics will, in the next life, be birds.

Plato

The Big Picture: In this chapter, we discuss the cosmic microwave background (CMB) radiation, the "snapshot" of the Universe at its infancy – when it was only about a few hundred thousand years old. We present the spectrum of the anisotropies of the CMB radiation, analyze its main features and consider their origins.

14.1. Importance of the CMB radiation

The CMB radiation is a prediction of the Big Bang theory. According to the theory, the early Universe was made up of a hot plasma of photons, electrons, and baryons. The photons were constantly interacting with the plasma through Thomson scattering. As the Universe expanded, adiabatic cooling caused the plasma to cool until it became favorable for electrons to combine with protons and form hydrogen atoms. This happened at around 3000 K or when the Universe was approximately 380,000 years old ($z \approx 1100$). At this point, the photons can no longer scatter off the now-neutral atoms and began to travel freely through space. This process is called *recombination* or *decoupling* (referring to electrons combining with nuclei and to the decoupling of matter and radiation, respectively).

The photons have continued cooling ever since; they have now reached 2.73 K, corresponding to about mm wavelength, placing it in the microwave domain. Their temperature will continue to drop as long as the Universe continues expanding ($T_\gamma \propto a^{-1}$). Accordingly, the microwave radiation from the sky that we measure today comes from a spherical surface, called the *surface of last scattering*. This represents the collection of points in space (currently around 46 billion light years from the Earth) at which the decoupling event happened long enough ago (less than 400,000 years after the Big Bang, 13.78 billion years ago) that the light from that part of space is just reaching us.

The Big Bang theory suggests that the CMB radiation fills all of observable space, and that most of the radiation energy in the Universe is in the CMB, which makes up a fraction of roughly 5×10^{-5} of the total density of the Universe today.

Relativity and Cosmology
https://doi.org/10.1016/B978-0-44-323542-9.00022-5

Two of the greatest successes of the Big Bang theory are its prediction of its almost perfect blackbody spectrum and its detailed prediction of the anisotropies in the CMB radiation. The WMAP and Planck CMB probes have precisely measured these anisotropies over the whole sky down to angular scales of 0.2 (13 arcminutes) and 0.08 degrees (5 arcminutes), respectively. These can be used to estimate the parameters of the standard ΛCDM model of the Big Bang, such as the shape of the Universe, its matter content, and the Hubble parameter.

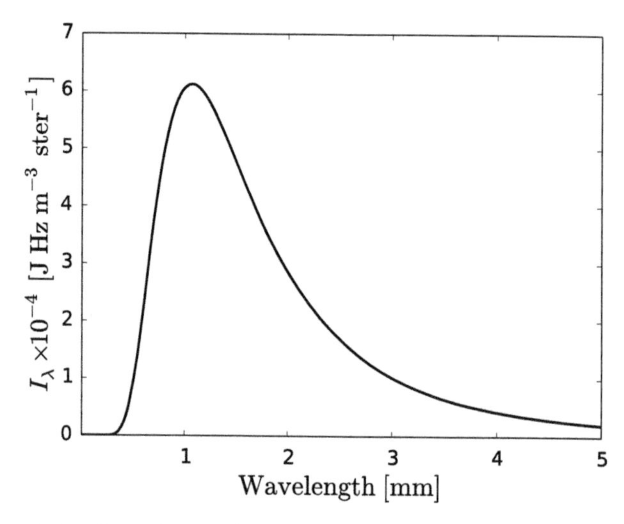

Figure 14.1 Intensity I_λ of the CMB radiation has a near-perfect blackbody spectrum, which peaks around $\lambda = 1$ mm. The differences between the theoretical prediction and the measured values are all much smaller than the thickness of the line.

14.1.1 Blackbody spectrum

The function describing the distribution of photons radiated by a blackbody is simply given by the equilibrium Bose–Einstein statistics, after taking $E = cp = h\nu = 2\pi\hbar\nu$, (where, recall, $h = 2\pi\hbar$ is the Planck constant):

$$f(\nu) = \frac{1}{e^{2\pi\hbar\nu/(kT)} - 1}, \tag{14.1}$$

and the corresponding intensity I of the blackbody spectrum is given by the *Planck distribution*:

$$I_\nu(\nu) \equiv \frac{dI}{d\nu} = \frac{4\pi\hbar}{c^2} \frac{\nu^3}{e^{2\pi\hbar\nu/(kT)} - 1}. \tag{14.2}$$

In terms of the wavelength λ, these are, after using $\lambda = c/\nu$:

$$f(\lambda) = \frac{c}{\lambda^2} \frac{1}{e^{2\pi\hbar c/(\lambda kT)} - 1},$$

$$I_\lambda(\lambda) \equiv \frac{dI}{d\lambda} = \frac{4\pi\hbar c^2}{\lambda^5} \frac{1}{e^{2\pi\hbar c/(\lambda kT)} - 1}. \tag{14.3}$$

The spectrum of the CMB radiation from Eq. (14.3), for $T_{\gamma,0} = 2.73$ K is shown in Fig. 14.1. The observations, first from COBE sattelite, followed by WMAP and then Planck telescopes agree with the perfect blackbody spectrum from Eq. (14.3) to within a few parts in 100,000, which means that the data points would be indistinguishable from the curve in Fig. 14.1.

14.1.2 Number density of the CMB photons

The energy density of the CMB photons is given by Eq. (12.21) as

$$u_\gamma = a_r T_\gamma^4. \tag{14.4}$$

The heat capacity associated with this blackbody radiation satisfies

$$C_\gamma \equiv \frac{du_\gamma}{dT_\gamma} = 4a_r T_\gamma^3. \tag{14.5}$$

Similarly if, in a first approximation, ordinary baryonic matter is assumed to consist completely of non-relativistic atoms not tied up in molecules, the heat capacity associated with baryonic matter satisfies

$$C_{\mathrm{m,b}} = \frac{3}{2} n_{\mathrm{b}} k = 4a_r T^3, \tag{14.6}$$

where n_{b} represents the number density of baryons. As we discussed earlier, it is not clear from observations what exactly the density of baryonic matter actually is. However, one can choose to parametrize one's ignorance by expressing the density in units of

$$\Omega = \frac{8\pi G\rho_0}{3H_0^2}, \tag{14.7}$$

where $\Omega = 1$ corresponds to the flat Universe. Given this parameterization, one infers that the ratio

$$\frac{C_{\mathrm{m,b}}}{C_\gamma} = \frac{3n_{\mathrm{b}}k}{4a_r T_\gamma^3} \approx 4 \times 10^{-9} \Omega h^2. \tag{14.8}$$

For any plausible values of Ω and h, this ratio is extremely small. This implies that, if photons and baryons were able to interact substantially so as to be in a thermal equilibrium, the baryons would be coupled completely to the photons. Given that the number density of baryons decreases as a^{-3}, *i.e.*, $n_b \sim a^{-3}$, it is also evident that, for the case of free-streaming photons, the ratio $C_{m,b}/C_\gamma$ is time independent. It follows that if photons and baryons were strongly coupled at some earlier epoch when the Universe was optically thick, it must have been the photons that regulated the resulting equilibrium. *The signatures of that pre-recombination equilibrium are imprinted into the CMB.*

Similarly, one can compare the rest mass energy of baryons with the energy associated with the blackbody radiation:

$$\frac{\rho_b c^2}{u_\gamma} = \frac{\rho_b c^2}{a_r T_\gamma^4} = \frac{\rho_{b,0} c^2 a^{-3}}{a_r T_{\gamma,0}^4 a^{-4}} = \frac{\rho_{b,0} c^2}{a_r T_{\gamma,0}^4} a \approx 4 \times 10^4 \Omega^2 h^2 a. \tag{14.9}$$

At the present time, the photon energy is negligible compared with the rest mass energy of the baryons. However, this photon energy was important dynamically at earlier times where $z \gtrsim 1000$.

The number density of the CMB photons n_γ can be roughly estimated as

$$n_\gamma = n_{\gamma,0} a^{-3} \approx \frac{\rho_{\gamma,0}}{\bar{E}_{\gamma,0}} a^{-3} = \frac{a_r T_{\gamma,0}^4}{3.5 k T_{\gamma,0}} a^{-3} \approx 420\ a^{-3}\ \text{cm}^{-3} = 420\ (1+z)^3\ \text{cm}^{-3}, \tag{14.10}$$

where we used $\bar{E}_{\gamma,0} \approx 3.5 k T_{\gamma,0}$ (left for the exercises). After recognizing

$$\bar{E}_\gamma = \bar{E}_{\gamma,0} a^{-1}, \tag{14.11}$$

we find that the energy density of the CMB is

$$u_\gamma = \bar{E}_\gamma n_\gamma = \left(\bar{E}_{\gamma,0} a^{-1}\right) 420\ a^{-3}\ \text{cm}^{-3} = 420\ \bar{E}_{\gamma,0} a^{-4}\ \text{cm}^{-3} \approx 0.26\ a^{-4}\ \text{eV cm}^{-3}. \tag{14.12}$$

14.2. Systematic bias: the dipole anisotropy

If the CMB radiation looks like a perfect blackbody radiation to one observer, it should not look like a perfect blackbody to other observers who are moving relative to the first observer. The radiation should be Doppler shifted because of the observer's motion. The observed radiation should appear somewhat bluer (hotter) in the direction in which the observer is moving and somewhat redder (cooler) in the opposite direction. The relativistic Doppler effects due to the motion of our frame of reference in relation to the frame of reference in which the CMB radiation is a perfect blackbody need to be accounted for before one can successfully analyze the CMB spectrum.

14.2.1 Relativistic Doppler shift

Assume the two observers are moving *away* from each other with a relative velocity v. Let us derive the special-relativity relation connecting the frequencies of light emitted in one (denoted with subscript 1) and received in another reference system (subscript 2), moving away at speed v.

Suppose one wavefront arrives at the observer. The next wavefront is then a distance $\lambda = c/\nu_1$ away from them (where λ is the wavelength, ν_1 the frequency of the wave emitted, and c is the speed of light). Since the wavefront moves with velocity c and the observer moves away with velocity v, the time observed between crests is, after using $\nu_1 = c/\lambda$,

$$t_1 = \frac{\lambda}{c - v} = \frac{\lambda}{\lambda\left(\frac{c}{\lambda} - \frac{v}{\lambda}\right)} = \frac{1}{\frac{c}{\lambda} - \frac{v}{c}\frac{c}{\lambda}} = \frac{1}{\left(1 - \frac{v}{c}\right)\nu_1}. \tag{14.13}$$

Due to the relativistic time dilation, the observer will measure this time to be

$$t_2 = \frac{t_1}{\gamma} = \frac{1}{\gamma\left(1 - \frac{v}{c}\right)\nu_1}, \tag{14.14}$$

where $\gamma = 1/\sqrt{1 - v^2/c^2}$, so the observed frequency is

$$\nu_2 = \frac{1}{t_2} = \gamma\left(1 - \frac{v}{c}\right)\nu_1, \tag{14.15}$$

and the corresponding relativistic Doppler shift:

$$\frac{\nu_2}{\nu_1} = \gamma\left(1 - \frac{v}{c}\right) = \frac{1 - \frac{v}{c}}{\sqrt{1 - \frac{v^2}{c^2}}}. \tag{14.16}$$

In a more general case, when the motion of the two reference frames is given by a vector $\hat{\boldsymbol{n}}$, such that $\mathbf{v}\hat{\boldsymbol{n}} = v\cos\theta$, the equation for the relativistic Doppler shift becomes

$$\frac{\nu_2}{\nu_1} = \frac{1 - \frac{\mathbf{v}\hat{\boldsymbol{n}}}{c}}{\sqrt{1 - \frac{v^2}{c^2}}} = \frac{1 - \frac{v}{c}\cos\theta}{\sqrt{1 - \frac{v^2}{c^2}}}. \tag{14.17}$$

However, *we* are moving in relation to the reference frame in which the CMB is a near-perfect blackbody, so we are $\nu_1 \equiv \nu_o$ and observing light that in the reference frame has frequency $\nu_2 \equiv \nu_e$, hence,

$$\frac{\nu_o}{\nu_e} = \frac{\sqrt{1 - \frac{v^2}{c^2}}}{1 - \frac{v}{c}\cos\theta}. \tag{14.18}$$

The temperature observed in the direction θ, $T(\theta)$, is given in terms of the average temperature $\langle T \rangle$:

$$\frac{T(\theta)}{\langle T \rangle} = \frac{\nu_o}{\nu_e} = \frac{\sqrt{1-\frac{v^2}{c^2}}}{1-\frac{v}{c}\cos\theta} = \left(1-\frac{v^2}{c^2}\right)^{1/2}\left(1-\frac{v}{c}\cos\theta\right)^{-1}$$
$$\approx \left(1-\frac{1}{2}\frac{v^2}{c^2}+...\right)\left(1+\frac{v}{c}\cos\theta+\frac{v^2}{c^2}\cos^2\theta+...\right)$$
$$\approx 1+\frac{v}{c}\cos\theta+\frac{v^2}{c^2}\left(\cos^2\theta-\frac{1}{2}\right)+.... \tag{14.19}$$

The motion of the observer (us) gives rise to both a dipole and other, higher-order corrections. The observed dipole anisotropy, first detected in the 1960s, implies that

$$\vec{v}_\odot - \vec{v}_{\mathrm{CMB}} = 370 \pm 10 \text{ km/s} \qquad \text{towards} \qquad \phi = 267.7 \pm 0.8°, \qquad \theta = 48.2 \pm 0.5°, \tag{14.20}$$

where θ is the *colatitude* (polar angle) and it is in the range $0 \le \theta \le \pi$ and ϕ is the *longitude* (azimuth) and it is in the range $0 \le \phi \le 2\pi$. Therefore $\theta = 0$ at the North Pole, $\theta = \pi/2$ at the Equator and $\theta = \pi$ at the South Pole.

Figure 14.2 CMB radiation temperature fluctuations from the 2016 Planck data seen over the full sky [52]. The average temperature is 2.73 K, and the colors represents small temperature fluctuations. Red regions are warmer, and blue colder by about 0.0002 K.

Allowing for the Sun's motion in the galaxy and the motion of the galaxy within the Local Group, this implies that the Local Group is moving with

$$\vec{v}_{\mathrm{LG}} - \vec{v}_{\mathrm{CMB}} \approx 600 \text{ km/s} \qquad \text{toward} \qquad \phi = 268°, \qquad \theta = 27°. \tag{14.21}$$

This *peculiar* motion[1] is subtracted from the measured CMB radiation, after which the intrinsic anisotropies are isolated (Fig. 14.2), and revealed to be about a few parts in 10^5. Even though minuscule, these primordial perturbations provided the seeds for the structure of the Universe.

14.3. Angular power spectrum

We now describe the approach that quantifies small-scale fluctuations in the CMB radiation field.

Define the *normalized deviation from the average temperature* $\langle T\rangle$ in direction $\hat{\boldsymbol{n}} \equiv (\theta, \phi)$ on the celestial sphere:

$$\Theta(\hat{\boldsymbol{n}}) \equiv \frac{T - \langle T\rangle}{\langle T\rangle} \equiv \frac{\Delta T}{\langle T\rangle}. \tag{14.22}$$

Next, we consider *multipole decomposition* of $\Theta(\hat{\boldsymbol{n}})$ in terms of *spherical harmonics* Y_{lm}:

$$\Theta(\hat{\boldsymbol{n}}) = \Theta(\theta, \phi) = \sum_{l=0}^{\infty} \sum_{m=-l}^{l} \Theta_{lm} Y_{lm}(\theta, \phi), \tag{14.23}$$

with the constant coefficients

$$\Theta_{lm} = \int \Theta(\hat{\boldsymbol{n}}) Y^*_{lm}(\hat{\boldsymbol{n}}) d\Omega. \tag{14.24}$$

The integral above is over the entire sphere and

$$Y_{lm}(\hat{\boldsymbol{n}}) = Y_{lm}(\theta, \phi) = \sqrt{\frac{(2l+1)}{4\pi} \frac{(l-m)!}{(l+m)!}} P_l^m(\cos\theta) e^{im\phi}. \tag{14.25}$$

$P_l^m(x)$ are the *associated Legendre functions*:

$$P_l^m(x) \equiv \frac{(1-x^2)^{m/2}}{2^l l!} \frac{d^{m+l}}{dx^{m+l}} \left(x^2 - 1\right)^l. \tag{14.26}$$

l is called the *multipole*, and characterizes an angular scale in the sky $\theta \approx 180°/l$.

[1] Peculiar motion is defined as motion with respect to the frame in which CMB is a near-perfect blackbody. The observers that measure CMB as a near-perfect blackbody are said to be a part of the *Hubble flow*.

The basis functions are orthonormal:

$$\int_{d\Omega} Y_{lm} Y^*_{l'm'} d\Omega = \int_{\theta=0}^{\pi} \int_{\phi=0}^{2\pi} Y_{lm} Y^*_{l'm'} \sin\theta \, d\phi \, d\theta = \delta_{ll'} \delta_{mm'}, \tag{14.27}$$

where $\delta_{nn'}$ is the Kronecker delta function ($\delta_{nn'} = 1$ when $n = n'$; $\delta_{nn'} = 0$ otherwise), and $d\Omega = \sin\theta \, d\phi \, d\theta$.

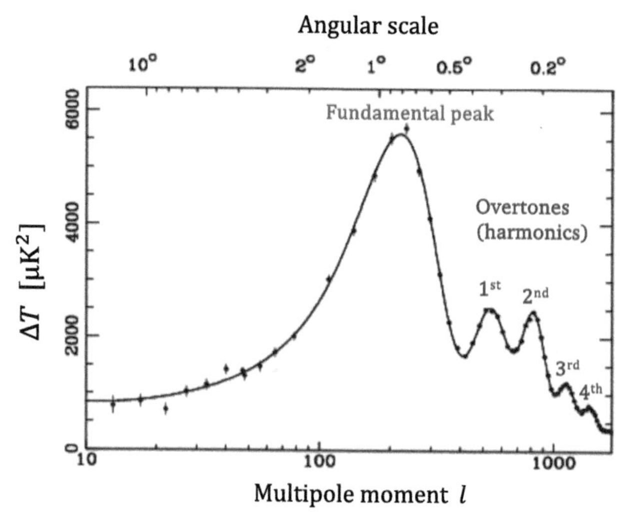

Figure 14.3 Power spectrum of the CMB radiation temperature anisotropies from observations. The ΛCDM fit is shown as a solid line. The upper axis shows the angular scale and the bottom axis the corresponding spherical harmonic multipoles.

The sum in Eq. (14.23) generally starts from $l = 2$ and extends to some $l_{\max}$ determined by the resolution of the data. The first (monopole) term, $l = 0$, is excluded because it is simply the average temperature fluctuation over the whole sky $\langle T(\hat{\mathbf{n}})\rangle$, should, by definition, be zero (left for the exercises). The second (dipole) term, $l = 1$ (corresponding to an angular size of $\theta \approx 180°$), is affected by our own motion through space (see Section 14.2 above). Our peculiar motion creates a dipole anisotropy that is considerably larger than the systematic cosmological dipole. Since there is no way for us to distinguish between the two, this contribution is usually not reported in the power spectrum.

The field of Gaussian random fluctuations is fully characterized by its power spectrum $\Theta^*_{lm}\Theta_{l'm'}$. The order m describes the angular orientation of a fluctuation mode, and the degree (multipole) l determines its characteristic angular size. Therefore in a Universe with no preferred direction (isotropic), we expect the power spectrum to be independent of m. Also, in a Universe that is the same from point to point (homogeneous), we expect the power spectrum to be independent of l.

14.3.1 Two-point correlation function vs. angular power spectrum

The power spectrum is related to the expected value of the correlation of temperature fluctuations between two points in the sky:

$$C(\theta) = C(\hat{\boldsymbol{n}} \cdot \hat{\boldsymbol{n}}') \equiv \langle \Theta(\hat{\boldsymbol{n}})\Theta(\hat{\boldsymbol{n}}') \rangle = \frac{1}{4\pi} \sum_{l=0}^{\infty} (2l+1) C_l P_l(\cos\theta), \tag{14.28}$$

where $P_l(\cos\theta)$ are the Legendre polynomials.

Here, we assume statistical isotropy, since the correlation function only depends on the angular separation between the two points θ. Therefore the power spectrum and the correlation function are defined by coefficients C_l, which only have a single coefficient, and not four, as in $C_{ll'mm'}$.

It can be shown (left for the exercises) that the angular power spectrum C_l is

$$\langle \Theta^*_{lm} \Theta_{l'm'} \rangle = \delta_{ll'} \delta_{mm'} C_l. \tag{14.29}$$

This relation implies that the coefficients of the angular power spectrum Θ_{lm} are uncorrelated.

It also states (when $l = l'$ and $m = m'$) that the variance in each coefficient is given by C_l defined in Eq. (14.28). It is important to reiterate that, because of rotational invariance, the indices corresponding to azimuthal angles ϕ and ϕ' are not a part of the two-point function amplitude C_l.

14.3.2 Cosmic variance

From Eq. (14.23), it follows that each of the multipoles l is determined by harmonics with $m \in [-l, l]$, a total of $2l+1$ samples from the same distribution. For example, while $l = 200$ may sample a respectable 401 coefficients, a quadrupole ($l = 2$) only samples 5. This poses a fundamental limit in determining the power, which is most restrictive for small l. This is called the *cosmic variance*:

$$\frac{\Delta C_l}{C_l} = \sqrt{\frac{2}{2l+1}}. \tag{14.30}$$

On small sections of the sky, if one zooms in far enough,[2] the spherical harmonic analysis becomes ordinary Fourier analysis in two dimensions. In this limit l becomes the

[2] That is, when a small-angle approximation is justified: $\sin\theta \approx \theta$, $\cos\theta \approx 1$, $\sin\phi \approx \phi$, $\cos\phi \approx 1$.

Fourier wavenumber. Since the angular wavelength $\theta = 2\pi/l$, large multipole moments correspond to small angular scales with $l \sim 10^2$ representing degree-scale separations.

The power spectrum is traditionally displayed in the literature as the power per logarithmic interval in l:

$$\Delta T^2 \equiv \frac{l(l+1)}{2\pi} C_l \langle T \rangle^2. \tag{14.31}$$

Fig. 14.3 shows the measurement of this quantity. The power spectrum shown in Fig. 14.3 begins at $l = 2$ and exhibits large errors at low multipoles due to cosmic variance.

Cosmic variance is particularly problematic for the monopole and dipole ($l = 0, 1$). If the monopole were larger in our vicinity than its average value, we would have no way of knowing it. Likewise for the dipole, we have no way of distinguishing a cosmological dipole from our own peculiar motion with respect to the CMB rest frame.

14.4. Scales in the angular power spectrum

The angular power spectrum quantifies the correlation of different parts of the sky we observe separated by an angle θ. This angle is related to a multipole l of the expansion as $\theta = 180°/l$. The size of the observable Universe (particle horizon) at the time of recombination (h_{rec} in Fig. 14.4) corresponds to about 1° ($l \approx 200$) on the sky today (ct_0, and particle horizon h_0 in Fig. 14.4). The part of the angular spectrum that correlates portions on the sky separated by angles appreciably larger than the size of the horizon at recombination (corresponding to $l \lesssim 20$) represent initial conditions: these parts of the Universe have not been in causal contact since before inflation (Fig. 14.4). The other part of the angular spectrum – at high l values – feature peaks corresponding to *acoustic oscillations* (Fig. 14.3). The positions and magnitudes of the peaks of acoustic oscillations contain fundamental properties about the geometry and structure of the Universe.

14.5. Baryon acoustic oscillations

In the early Universe before recombination, rapid scattering couples photons, electrons, and baryons into a plasma that behaves as perfect fluid, well described by the $P = \rho c^2/3$ equation of state. Also mixed in is non-baryonic, pressureless dark matter, which shares the density inhomogeneities of the baryon–photon plasma. Initial quantum overdensities create gravitational potential wells – inflationary seeds of the Universe's structure. Infall of the electrons and baryons into the potential wells is resisted by its pressure (recall

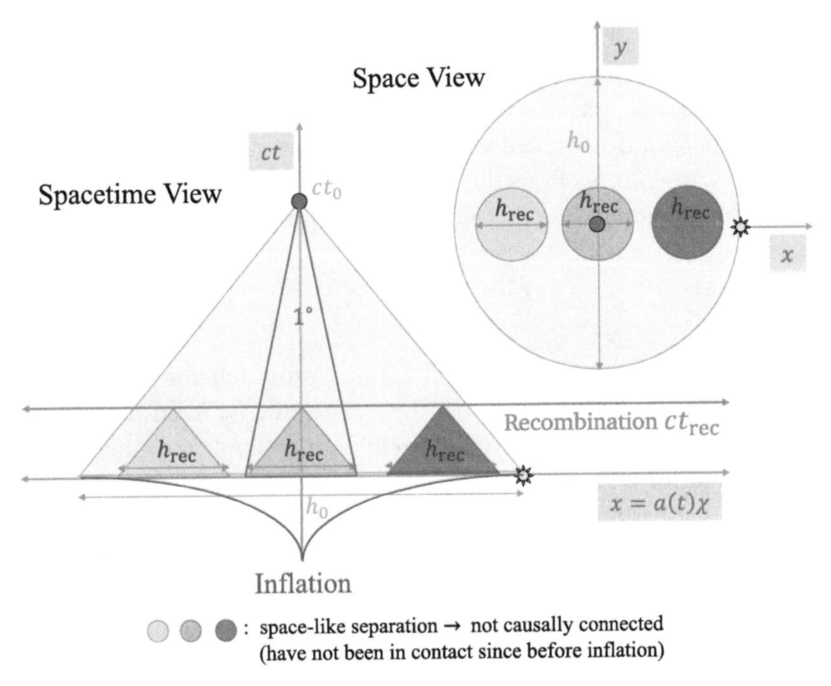

Figure 14.4 CMB particle horizon after the inflation (or another mechanism that puts matter initially in causal contact, as outlined in Section 9.5). The star is just entering the particle horizon for the observer located at the dot along the *ct* axis at ct_0. Both spacetime (bottom left) and space (top right) views are shown. The three shaded regions are remarkably similar, which strongly suggests that at one point in time they were in causal contact and in equilibrium. Exponential inflation ensures such causal contact.

$P = \rho c^2/3$), thus forming *acoustic oscillations*[3]: periodic compressions (overdensities in the fluid) and bounce (underdensities). These acoustic oscillations of the early Universe are frozen at recombination and give the CMB spectrum a unique signature.

The CMB data reveal that the initial inhomogeneities in the Universe were small. Overdense regions would grow by gravitationally attracting more mass, but only after the entire region is in causal contact. This means that only regions that are smaller than the horizon at recombination had time to compress before then. Regions that are sufficiently smaller than the horizon had enough time to compress gravitationally until the outward-acting pressure halted the compression via Thomson scattering, and possibly even go through a number of such acoustic oscillations. Therefore perturbations of particular sizes may have gone through: (1) one compression (fundamental wave);

[3] Acoustic oscillation is just another name for a density wave. The speed of sound in a medium determines the speed at which density oscillations/perturbations/waves travel.

(2) one compression and one bounce (first overtone); (3) one compression, one bounce, and one compression again (second overtone); etc. (these are the peaks in Fig. 14.3).

At the end of the inflationary era, the acoustic oscillations start everywhere in phase. The longest wavelength is associated with the perturbation that has gone through one-half of the density oscillation from the end of inflation (when the oscillation started, $t = 0$) until recombination (when the oscillations stopped, $t = t_{\text{rec}}$):

$$\lambda_1 = 2v_s t_{\text{rec}} = \frac{2ct_{\text{rec}}}{\sqrt{3}}, \tag{14.32}$$

where $v_s \approx c/\sqrt{3}$ is the speed of sound in plasma (proof left for the exercises). This one-half of a density oscillation was either from an overdense to an underdense region (bounce) or the other way around (compression). In a similar manner, higher-order modes are achieved for regions that went through multiple cycles between overdense and underdense: first harmonic has a wavelength $\lambda_2 = 2\lambda_1$; second harmonic $\lambda_3 = 3\lambda_1$, etc. At the end of recombination, the baryonic matter and photons decouple, and these overdense (cold) and underdense (hot) regions leave a permanent imprint in the CMB radiation field.[4] These are the spots of the surveyed CMB sky where temperatures are different from the mean. The largest deviation from the mean (hottest/coldest spots) will be due to those acoustic oscillations that were at their extremum (hottest/coldest) by the end of recombination: the fundamental and the overtones. Therefore the CMB spectrum should have peaks at these specific scales.

Consider a standing wave $A_k(x, t) \propto \sin(kx)\cos(\omega t)$, going through space at the speed of sound, with the frequency ω and wavenumber k, related by $\omega = kv_s$. The displacement – and hence the correlation in temperature – will be extremal at the recombination time t_{rec} for $\omega t_{\text{rec}} = kv_s t_{\text{rec}} = \pi, 2\pi, 3\pi....$ The subsequent peaks in the power spectrum represent the temperature variations caused by overtones. If the perturbations had been continuously generated over time, the power spectrum would not be so harmoniously ordered.

14.5.1 Dampening of the overtones

Both ordinary matter and dark matter supply mass to the primordial plasma and enhance the gravitational pull, but only ordinary matter undergoes the sonic compressions and bounces (the pressureless dark matter has decoupled from the plasma at a much earlier time). At recombination, the fundamental wave is frozen in a phase where gravity enhances its compression of the denser regions of plasma. The first overtone, which corresponds to scales half of the fundamental wavelength, is caught in the opposite phase

[4] The CMB photons originating from the overdense regions lost a portion of their energy "climbing out" of the gravitational well, thereby are less energetic ("colder") than those emerging from underdense regions.

– gravity is attempting to compress the plasma while the plasma pressure is trying to expand it. As a consequence, the temperature variations caused by this overtone (and all subsequent ones) will be less pronounced than those caused by the fundamental wave (fundamental peak) [53]. This dampening of the magnitudes of the overtones allows for quantification of the relative strength of gravity and radiation pressure in the early Universe.

The theory of inflation predicts that the sound waves should have nearly the same amplitude on all scales. The power spectrum, however, shows a sharp drop-off in magnitude of temperature variations after the third peak. This is due to the dissipation of the sound waves with short wavelengths: sound is carried by oscillation of particles in gas or plasma, so a wave cannot propagate if its wavelength is shorter than the typical distance traveled by particles between collisions.

14.6. Physical effects affecting the CMB radiation

14.6.1 Sunyaev–Zel'dovich effect

The Sunyaev–Zel'dovich (SZ) effect refers to the inverse Compton scattering[5] of the CMB photons by high-energy electrons from the hot, ionized gas in clusters of galaxies. It was first predicted in 1969 by Sunyaev and Zel'dovich [27]. The effect is a *secondary anisotropy* to the CMB.[6] The SZ effect causes a "hotspot" in the CMB due to the kinetic SZ effect (the bulk motion of the cluster with respect to the CMB) and a noticeable change in the shape of the CMB spectrum due to the thermal SZ effect.

The SZ effect is important to the study of cosmology and the CMB for two main reasons:

1. The observed "hotspots" created by the kinetic effect will distort the power spectrum of the CMB anisotropies. These need to be separated from the primary anisotropies in order to probe the properties of inflation.
2. The thermal SZ effect can be measured and combined with X-ray observations in order to determine values of cosmological parameters, in particular the present value of the Hubble parameter, H_0.

14.6.2 Sachs–Wolfe effect

At last scattering, the baryonic matter and photons decouple and the photons suddenly find themselves free to travel in straight paths through the Universe. However, the

[5] *Inverse* Compton scattering is scattering of photons off high-energy electrons in which energy/momentum transfer is from "hot" electrons onto photons. In "ordinary" Compton scattering, the transfer goes the other way – from highly energetic photons to electrons.

[6] Here, "secondary anisotropy" means that it took place since recombination, as the photons were traveling toward us. These type of effects are also referred to as *foreground anisotropies*.

baryons are clustered together in gravitational potential wells prior to last scattering. Since the photons are tightly coupled to the baryons before last scattering, they were initially confined to these gravitational potential wells too. Thus the photons have to climb out of gravitational potential wells when they are suddenly freed at last scattering. The climb costs photons some energy, therefore making them gravitationally redshifted. This effect is known as the *Sachs–Wolfe* (SW) effect. Since it is imprinted on the CMB power spectrum at the time of last scattering, it is considered a *primary anisotropy*.

This SW effect is the principal source of fluctuations in the CMB for angular scales above about ten degrees – the regions in the early Universe that were too big to undergo acoustic oscillations.

14.6.3 Integrated Sachs–Wolfe effect

The integrated Sachs–Wolfe (ISW) effect is also due to gravitational redshift, but here it takes place between the surface of last scattering and the Earth, so it is not a fundamental part of the CMB.

The ISW effect can arise after last scattering as the photons free stream through the Universe. Although the photons are no longer tightly coupled to baryonic matter, they can still slip into gravitational potential wells and have to climb back out. When they fall in, the photons gain some energy (are blueshifted), and when they climb back out, they lose some (are redshifted). Assuming that the depth of the gravitational potential well remains constant while the photon traverses it, the redshift exactly cancels the blueshift. No trace of the photon's passage through the gravitational potential well remains, assuming that both sides of the dip are the same height and no energy is dissipated. Suppose, however, that the gravitational potential well through which the photon passes either decays or deepens while the photon is inside. Then, its redshift and blueshift will not exactly cancel; instead the photon gains or loses some energy (respectively) from its passage through the gravitational potential well.

There are two main contributions to the ISW. The first occurs shortly after photons leave the last-scattering surface, and is due to the evolution of the gravitational potential wells as the Universe changes from being dominated by radiation to being dominated by matter. The second, sometimes called the *late-time integrated Sachs–Wolfe effect*, arises much later as the evolution starts to feel the effect of the cosmological constant (or, more generally, dark energy), or curvature of the Universe if it is not flat. The latter effect has an observational signature in the amplitude of the large-scale perturbations of the CMB and their correlation with the large-scale structure.

The primary anisotropies (SW) on the CMB power spectrum tell us about the initial conditions of the photons at the time of last scattering. The secondary anisotropies (ISW, SZ) carry information about a change in these initial conditions due to the voyage of the CMB photons from the surface of last scattering to our instruments today.

Exercises

1. Show that the blackbody distribution of the CMB at the time recombination retains its blackbody form as the Universe expands.
2. Show that $\Theta(\hat{\boldsymbol{n}})_{l_{\max}=0} = \langle \Theta(\hat{\boldsymbol{n}}) \rangle = 0$.
3. Derive Eq. (14.29). Hint: Use the addition theorem:

$$P_l(\hat{\boldsymbol{n}} \cdot \hat{\boldsymbol{n}}') = \frac{4\pi}{2l+1} \sum_{m=-l}^{l} Y_{lm}(\hat{\boldsymbol{n}}) Y^*_{lm}(\hat{\boldsymbol{n}}'). \tag{14.33}$$

4. Show that a half of the CMB photons observed now have energies below $3.5kT_{\gamma,0}$.
5. Show that the speed of sound in plasma is $v_s = c/\sqrt{3}$.
6. Derive Eq. (14.3) from Eq. (14.2).

Epilogue

Throughout my life, I have encountered many passionate admirers of astrophysics and cosmology, not only young students in the classroom but also professionals from all walks of life – curious lawyers, retired engineers, a tipsy philosopher... Inspired by the master science proselytizers like Carl Sagan or Neil deGrasse Tyson, wide-eyed and visibly excited, they would pepper me with questions about black holes, wormholes, dark matter, and dark energy. As I would answer to the best of my ability, one thing would inevitably become clear to my interlocutors: a deeper understanding of this engrossing subject requires *mathematical rigor.* The path from dazzling storytelling to the fundamental understanding is paved with equations!

The primary purpose of this book is to transition from a qualitative, layperson's description toward a more profound, quantitative understanding of cosmology. The mathematics used to accomplish this is at the level of elementary calculus (integration, differentiation, and an occasional ordinary differential equation). Beyond that, just enough tensor calculus (and not an index more!) was introduced to derive Einstein's field equations that quantify how gravity works.

The book's secondary purpose is to embolden its readers to continue their study of cosmology. It is my hope that this introductory journey into general relativity and modern cosmology leaves them wanting more. Certainly, there is more, much more, that did not make it into this stand-alone book covering exactly one college semester's worth of introductory material. This includes details of topics such as Big Bang nucleosynthesis, the dark ages, reionization, structure formation and evolution, new revelations that are beginning to emerge from James Webb Space Telescope, and others.

There are topics mentioned in this volume that are among the most intriguing mysteries of all sciences: dark matter and dark energy. The new generation of scientists learning about these subjects right now might finally solve these tantalizing puzzles. It is my sincere wish that some of those emerging scientists benefit from this book.

Further reading

1. **Relativity**. A good place to gently start on the subject is McMahon 2005 [54]. From there, it is good to continue with the standard text by Schutz [55]. A rigorous and technical yet clear treatment of the tensor calculus and relativity is provided by Dalarsson and Dalarsson 2005 [56]. Of course, the "bible" on the subject of tensor calculus, and really everything to do with relativity, is all meticulously derived and reported in the *magnum opus* by Misner, Thorne, and Wheeler 1973 [8]. This treatise is not for the faint-hearted!
2. **Cosmology**. An excellent first text in cosmology was written by Ryden [57], and has been one of the standards in the field for over two decades. A recent arrival to the scene, a particularly clear textbook with an emphasis on statistical and data analysis techniques by Huterer [58], is quickly becoming one of my favorites. For a very detailed, highly technical text, see Weinberg 2008 [59].
3. **Dark Matter**. A popular overview of the dark-matter searches is provided by Hooper 2006 [60]. An interesting historical account on dark matter was given by Bertone and Hooper 2016 [61], and a detailed review was given by Strigari 2013 [62]. For an outstanding comparison of the ΛCDM paradigm with modified Newtonian dynamics (MOND), see Merritt 2020 [63].
4. **Dark Energy**. The cosmological constant problem was discussed in detail in Weinberg 1989 [64]. The original Nobel prize-winning work is a must-read: Riess et al. 1998 [46] and Perlmutter et al. 1999 [45], as well as the Swedish Royal Academy's justification for awarding them the 2011 Nobel Prize in physics [47].
5. **CMB**. An excellent place to start reading about CMB is Wayne Hu's tutorials, which can be found online at [65]. It has tutorials at elementary, intermediate, and advanced levels. Outstanding illustrations of the various aspects of the CMB power spectrum, fundamental and overtones, and acoustic oscillations are found in a popular article by Hu and White 2004 [53]. For a very detailed, technical treatment, see Weinberg 2008 [59] and Dodelson 2003 [66].

APPENDIX A

An alternative Lagrangian

In Eq. (4.80) we used an alternative Lagrangian,

$$L = \frac{1}{2} g_{\gamma\delta} \dot{x}^{\gamma} \dot{x}^{\delta},$$

instead of the traditional

$$L = \sqrt{g_{\gamma\delta} \dot{x}^{\gamma} \dot{x}^{\delta}}.$$

Both of these should give the same result, because an extremal of a functional L will also extremize L^2 and vice versa. Here, we present a mathematical justification why either works correctly, *i.e.*, why the expression given in Eq. (4.80) is a Lagrangian that generates the geodesic equation.

We prove that by applying the Lagrange equation

$$\frac{\partial L}{\partial x^{\alpha}} - \frac{d}{d\lambda} \frac{\partial L}{\partial \dot{x}^{\alpha}} = 0$$

to the expression in Eq. (4.80), and recovering the geodesic equation:

$$\begin{aligned}
\frac{\partial L}{\partial x^{\alpha}} &= \frac{1}{2} g_{\gamma\delta,\alpha} \dot{x}^{\gamma} \dot{x}^{\delta}, \\
\frac{\partial L}{\partial \dot{x}^{\alpha}} &= \frac{1}{2} g_{\alpha\delta} \dot{x}^{\delta} + \frac{1}{2} g_{\gamma\alpha} \dot{x}^{\gamma}, \\
\frac{d}{d\lambda}\left(\frac{\partial L}{\partial \dot{x}^{\alpha}}\right) &= \frac{1}{2} g_{\alpha\delta,\gamma} \dot{x}^{\delta} \dot{x}^{\gamma} + \frac{1}{2} g_{\alpha\delta} \ddot{x}^{\delta} + \frac{1}{2} g_{\gamma\alpha,\delta} \dot{x}^{\gamma} \dot{x}^{\delta} + \frac{1}{2} g_{\gamma\alpha} \ddot{x}^{\gamma} \\
&= \frac{1}{2}\left(g_{\alpha\delta,\gamma} + g_{\gamma\alpha,\delta}\right) \dot{x}^{\delta} \dot{x}^{\gamma} + g_{\alpha\delta} \ddot{x}^{\delta},
\end{aligned}$$

as we are at liberty to rename dummy variables (ones that are summed over), and to exchange indices of the metric tensor, since it is symmetric. The Lagrange equation therefore reads:

$$\begin{aligned}
\frac{\partial L}{\partial x^{\alpha}} - \frac{d}{d\lambda} \frac{\partial L}{\partial \dot{x}^{\alpha}} &= \frac{1}{2} g_{\gamma\delta,\alpha} \dot{x}^{\gamma} \dot{x}^{\delta} - \frac{1}{2}\left(g_{\alpha\delta,\gamma} + g_{\gamma\alpha,\delta}\right) \dot{x}^{\delta} \dot{x}^{\gamma} - g_{\alpha\delta} \ddot{x}^{\delta} = \\
&= -\frac{1}{2}\left(g_{\alpha\delta,\gamma} + g_{\gamma\alpha,\delta} - g_{\gamma\delta,\alpha}\right) \dot{x}^{\delta} \dot{x}^{\gamma} - g_{\alpha\delta} \ddot{x}^{\delta} = 0.
\end{aligned}$$

Now, multiply both sides by $g^{\nu\alpha}$ to isolate the second derivative term:

$$\ddot{x}^{\nu} = -\frac{1}{2} g^{\nu\alpha}\left(g_{\alpha\delta,\gamma} + g_{\gamma\alpha,\delta} - g_{\gamma\delta,\alpha}\right) \dot{x}^{\delta} \dot{x}^{\gamma}.$$

However, by definition,

$$\Gamma^{\nu}_{\beta\gamma} = \frac{1}{2} g^{\nu\alpha} \left(g_{\alpha\delta,\gamma} + g_{\gamma\alpha,\delta} - g_{\gamma\delta,\alpha} \right),$$

hence, we finally have

$$\ddot{x}^{\nu} = -\Gamma^{\nu}_{\beta\gamma} \dot{x}^{\delta} \dot{x}^{\gamma},$$

which is the geodesic equation we derived in Eq. (4.87). This proves that the Lagrangian in Eq. (4.80) also generates the geodesic equation.

APPENDIX B

Geodesic equation in spherical coordinates

Let us compute the geodesic in 3D flat space, expressed in spherical coordinates $x^i = (x^1, x^2, x^3) = (r, \theta, \phi)$. This should be an analog to geodesics in flat space in Cartesian coordinates:

$$\ddot{x}^\alpha = 0.$$

This can be done in at least two ways.

Method 1: Brute force – computing Christoffel symbols and substituting them into the geodesic equation:

$$\ddot{x}^\nu = -\Gamma^\nu_{\beta\gamma}\dot{x}^\beta\dot{x}^\gamma. \tag{B.1}$$

From Eq. (4.68), the Christoffel symbols for the spherical space are given by

$$\Gamma^k_{ij} = \frac{1}{2}p^{kl}\left(p_{il,j} + p_{lj,i} - p_{ij,l}\right),$$

where

$$p_{ij} = \begin{pmatrix} 1 & 0 & 0 \\ 0 & r^2 & 0 \\ 0 & 0 & r^2\sin^2\theta \end{pmatrix}, \quad \text{and} \quad p^{ij} = \begin{pmatrix} 1 & 0 & 0 \\ 0 & \frac{1}{r^2} & 0 \\ 0 & 0 & \frac{1}{r^2\sin^2\theta} \end{pmatrix}.$$

Since $p_{11} = 1$, all of its derivatives vanish. Also, because of symmetry (look at the definition given in Eq. (4.68) and recall that the metric tensor is symmetric), we have

$$\Gamma^k_{1j} = \Gamma^k_{j1} = \frac{1}{2}p^{kl}\left(p_{1l,j} + p_{lj,1} - p_{1j,l}\right) = \frac{1}{2}p^{kl}p_{lj,1} = \frac{1}{2}\left(p^{k2}p_{2j,1} + p^{k3}p_{3j,1}\right)$$

and

$$\begin{aligned}
&\Gamma^1_{1j} = \Gamma^1_{j1} = 0, && \text{as } p_{12} = 0, p_{13} = 0, \\
&\Gamma^1_{22} = \frac{1}{2}p^{1l}\left(p_{2l,2} + p_{l2,2} - p_{22,l}\right) = -\frac{1}{2}p^{11}p_{22,1} = -r, && \\
&\Gamma^1_{23} = \Gamma^1_{32} = \frac{1}{2}p^{1l}\left(p_{2l,3} + p_{l3,2} - p_{23,l}\right) = \frac{1}{2}p^{1l}p_{l3,2} = 0, && \text{as } p_{23} = 0, p_{ij,3} = 0, \\
&\Gamma^1_{33} = \frac{1}{2}p^{1l}\left(p_{3l,3} + p_{l3,3} - p_{33,l}\right) = -\frac{1}{2}p^{11}p_{33,1} = -r\sin^2\theta, &&
\end{aligned}$$

$$\Gamma^2_{1j} = \frac{1}{2}p^{22}p_{2j,1},$$
$$\Gamma^2_{11} = 0,$$
$$\Gamma^2_{12} = \Gamma^2_{21} = \frac{1}{2}p^{22}p_{22,1} = \frac{1}{2}\frac{1}{r^2}2r = \frac{1}{r},$$
$$\Gamma^2_{13} = \Gamma^2_{31} = 0,$$
$$\Gamma^2_{22} = 0,$$
$$\Gamma^2_{23} = \Gamma^2_{32} = \frac{1}{2}p^{2l}\left(p_{2l,3} + p_{l3,2} - p_{23,l}\right) = \frac{1}{2}p^{22}\left(p_{22,3} + p_{23,2} - p_{23,2}\right) = 0,$$
$$\Gamma^2_{33} = \frac{1}{2}p^{2l}\left(p_{3l,3} + p_{l3,3} - p_{33,l}\right) = -\frac{1}{2}p^{22}p_{33,2} = -\frac{1}{2}\frac{1}{r^2}(2r^2\sin\theta\cos\theta) = -\sin\theta\cos\theta,$$
$$\Gamma^3_{ij} = \frac{1}{2}p^{3l}\left(p_{il,j} + p_{lj,i} - p_{ij,l}\right) = \frac{1}{2}p^{33}\left(p_{i3,j} + p_{3j,i} - p_{ij,3}\right) = \frac{1}{2}p^{33}\left(p_{i3,j} + p_{3j,i}\right),$$
$$\Gamma^3_{11} = 0,$$
$$\Gamma^3_{12} = \Gamma^3_{21} = 0,$$
$$\Gamma^3_{13} = \Gamma^3_{31} = \frac{1}{2}p^{33}\left(p_{13,3} + p_{33,1}\right) = \frac{1}{2}p^{33}p_{33,1} = \frac{1}{2}\frac{1}{r^2\sin^2\theta}(2r\sin^2\theta) = \frac{1}{r},$$
$$\Gamma^3_{22} = 0,$$
$$\Gamma^3_{23} = \Gamma^3_{32} = \frac{1}{2}p^{33}\left(p_{23,3} + p_{33,2}\right) = \frac{1}{2}p^{33}p_{33,2} = \frac{1}{2}\frac{1}{r^2\sin^2\theta}(2r^2\sin\theta\cos\theta) = \cot\theta,$$
$$\Gamma^3_{33} = \frac{1}{2}p^{33}\left(p_{33,3} + p_{33,3}\right) = 0.$$

The geodesic equation in spherical coordinates, given in Eq. (B.1), then becomes (recall $x^1 = r$, $x^2 = \theta$, $x^2 = \phi$)

$$\begin{aligned}
\ddot{x}^1 = \ddot{r} &= -\Gamma^1_{\gamma\delta}\dot{x}^\gamma\dot{x}^\delta = -\Gamma^1_{22}(\dot{x}^2)^2 - \Gamma^1_{33}(\dot{x}^3)^2 \\
&= r\dot{\theta}^2 + r\sin^2\theta\dot{\phi}^2, \\
\ddot{x}^2 = \ddot{\theta} &= -\Gamma^2_{\gamma\delta}\dot{x}^\gamma\dot{x}^\delta = -2\Gamma^2_{12}\dot{x}^1\dot{x}^2 - \Gamma^2_{33}(\dot{x}^3)^2 \\
&= -\frac{2}{r}\dot{r}\dot{\theta} + \sin\theta\cos\theta\dot{\phi}^2, \\
\ddot{x}^3 = \ddot{\phi} &= -\Gamma^3_{\gamma\delta}\dot{x}^\gamma\dot{x}^\delta = -2\Gamma^3_{13}\dot{x}^1\dot{x}^3 - 2\Gamma^3_{23}\dot{x}^2\dot{x}^3 \\
&= -\frac{2}{r}\dot{r}\dot{\phi} - 2\cot\theta\dot{\theta}\dot{\phi}.
\end{aligned}$$

Method 2: Using a Lagrangian $L = g_{\gamma\delta}\dot{x}^\gamma\dot{x}^\delta$. This alternative Lagrangian becomes

$$L = p_{ij}\dot{x}^i\dot{x}^j = \dot{r}^2 + r^2\dot{\theta}^2 + r^2\sin^2\theta\dot{\phi}^2,$$

hence, applying the Lagrange equations,

$$\frac{\partial L}{\partial x^l} - \frac{d}{d\lambda}\frac{\partial L}{\partial \dot{x}^l} = 0,$$

yields, for each coordinate r, θ, ϕ,

$$\frac{\partial L}{\partial r} - \frac{d}{d\lambda}\frac{\partial L}{\partial \dot{r}} = 2r\dot{\theta}^2 + 2r\sin^2\theta\dot{\phi}^2 - 2\ddot{r} = 0 \rightarrow \ddot{r} = r\dot{\theta}^2 + r\sin^2\theta\dot{\phi}^2$$
$$\frac{\partial L}{\partial \theta} - \frac{d}{d\lambda}\frac{\partial L}{\partial \dot{\theta}} = 2r^2\sin\theta\cos\theta\dot{\phi}^2 - 4r\dot{r}\dot{\theta} - 2r^2\ddot{\theta} = 0, \rightarrow \ddot{\theta} = -\frac{2}{r}\dot{r}\dot{\theta} + \sin\theta\cos\theta\dot{\phi}^2$$
$$\frac{\partial L}{\partial \phi} - \frac{d}{d\lambda}\frac{\partial L}{\partial \dot{\phi}} = -4r\dot{r}\sin^2\theta\dot{\phi} - 4r^2\sin\theta\cos\theta\dot{\theta}\dot{\phi} - 2r^2\sin^2\theta\ddot{\phi} = 0$$
$$\rightarrow \ddot{\phi} = -\frac{2}{r}\dot{r}\dot{\phi} - 2\cot\theta\dot{\theta}\dot{\phi}.$$

This set of equations represents motion in flat space, as described by spherical coordinates, and therefore should describe straight lines. This is fairly easy to see for purely radial motion in the $x-y$ plane, $\theta = \pi/2$ and $\phi = const.$, so the right-hand side of all three geodesic equations above vanish, and we recover a straight (radial) line $\ddot{r} = 0$. In a more general case, it is less trivial to show that the equations above represent straight lines.

Using this alternative Lagrangian allows one to readily read off Christoffel symbols, as we compare the differential equations above with the definition of a geodesic:

$$\ddot{x}^\nu = -\Gamma^\nu_{\beta\gamma}\dot{x}^\beta\dot{x}^\gamma.$$

From the equations above, they are readily identified as

$$\Gamma^1_{22} = -r,$$
$$\Gamma^1_{33} = -r\sin^2\theta,$$
$$\Gamma^2_{12} = \Gamma^2_{21} = \frac{1}{r},$$
$$\Gamma^2_{33} = -\sin\theta\cos\theta,$$
$$\Gamma^3_{13} = \Gamma^3_{31} = \frac{1}{r},$$
$$\Gamma^3_{23} = \Gamma^3_{32} = \cot\theta,$$

just as we computed by brute force. The factor 2 in front of Christoffel symbols Γ^i_{jk} that have unequal lower indices ($j \neq k$) reflects the fact that, because of symmetry, both Γ^i_{jk} and Γ^i_{kj} are counted.

It is not advisable to compute the geodesic equation from the traditional Lagrangian $L = \sqrt{g_{\gamma\delta}\dot{x}^\gamma\dot{x}^\delta}$, as it will quickly lead to some *extremely* cumbersome algebra. The three

Lagrangian equations should *eventually* reduce to the geodesic equations we derived above (because the two are equivalent in terms of producing the same result), but it quickly becomes obvious which approach is preferable.

APPENDIX C

Example of metric conversion

Let us see how to convert from one space metric to another, *i.e.*, use Eq. (4.23).

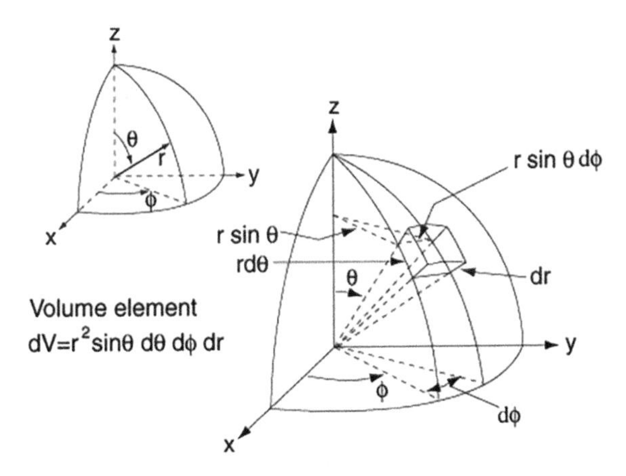

Figure C.1 Spherical coordinate system.

For example, given the space metric in Cartesian coordinates $(x^1, x^2, x^3) = (x, y, z)$ (shown in Fig. C.1),

$$\delta_{ij} = \begin{pmatrix} 1 & 0 & 0 \\ 0 & 1 & 0 \\ 0 & 0 & 1 \end{pmatrix},$$

let us find the space metric in spherical coordinates $(x'^1, x'^2, x'^3) = (r, \theta, \phi)$. Cartesian coordinates are given in terms of spherical coordinates as

$$x = r\sin\theta\cos\phi, \quad y = r\sin\theta\sin\phi, \quad z = r\cos\theta,$$

or

$$x^1 = x'^1 \sin x'^2 \cos x'^3, \quad x^2 = x'^1 \sin x'^2 \sin x'^3, \quad x^3 = x'^1 \cos x'^2.$$

Then,

$$\frac{\partial x^1}{\partial x'^1} = \sin x'^2 \cos x'^3, \quad \frac{\partial x^1}{\partial x'^2} = x'^1 \cos x'^2 \cos x'^3, \quad \frac{\partial x^1}{\partial x'^3} = -x'^1 \sin x'^2 \sin x'^3,$$
$$\frac{\partial x^2}{\partial x'^1} = \sin x'^2 \sin x'^3, \quad \frac{\partial x^2}{\partial x'^2} = x'^1 \cos x'^2 \sin x'^3, \quad \frac{\partial x^2}{\partial x'^3} = x'^1 \sin x'^2 \cos x'^3,$$

$$\frac{\partial x^3}{\partial x'^1} = \cos x'^2, \qquad \frac{\partial x^3}{\partial x'^2} = -x'^1 \sin x'^2, \qquad \frac{\partial x^3}{\partial x'^3} = 0.$$

From Eq. (4.23), we have

$$\begin{aligned}
ds^2 &= \delta_{ij} dx^i dx^j \\
&= \delta_{ij} \frac{\partial x^i}{\partial x'^k} \frac{\partial x^j}{\partial x'^l} dx'^k dx'^l \\
&= (dx'^1)^2 \left[\sin^2 x'^2 \cos^2 x'^3 + \sin^2 x'^2 \sin^2 x'^3 + \cos^2 x'^2\right] \\
&\quad + (dx'^2)^2 \left[(x'^1)^2 \cos^2 x'^2 \cos^2 x'^3 + (x'^1)^2 \cos^2 x'^2 \sin^2 x'^3 + (x'^1)^2 \sin x'^3\right] \\
&\quad + (dx'^3)^2 \left[(x'^1)^2 \sin^2 x'^2 \sin^2 x'^3 + (x'^1)^2 \sin^2 x'^2 \cos^2 x'^3\right] \\
&= (dx'^1)^2 + (x'^1)^2 (dx'^2)^2 + (x'^1)^2 \sin^2 x'^2 (dx'^3)^2 \\
&= dr^2 + r^2 d\theta^2 + r^2 \sin^2\theta \, d\phi^2 \\
&= p_{11}(dr)^2 + p_{22}(d\theta)^2 + p_{33}(d\phi)^2 \\
&= p_{11}(dx'^1)^2 + p_{22}(dx'^2)^2 + p_{33}(dx'^3)^2 = p_{ij} dx'^i dx'^j.
\end{aligned}$$

Reading off the diagonal components of the metric, we have

$$\begin{aligned}
p_{11} &= 1, \\
p_{22} &= r^2, \\
p_{33} &= r^2 \sin^2\theta,
\end{aligned}$$

hence, the space metric for spherical coordinates is

$$p_{ij} = \begin{pmatrix} 1 & 0 & 0 \\ 0 & r^2 & 0 \\ 0 & 0 & r^2 \sin^2\theta \end{pmatrix} \quad \text{or} \quad p^{ij} = \begin{pmatrix} 1 & 0 & 0 \\ 0 & \frac{1}{r^2} & 0 \\ 0 & 0 & \frac{1}{r^2 \sin^2\theta} \end{pmatrix}.$$

APPENDIX D

Applying the geodesic equation

Let us compute the geodesic equation on the surface of the 2-sphere embedded in a 3D space. The radius is then constant $r = R$, the coordinates are $(x^1, x^2) = (\theta, \phi)$, and the metric is

$$p_{ij} = \begin{pmatrix} R^2 & 0 \\ 0 & R^2 \sin^2\theta \end{pmatrix}.$$

The Lagrangian again is

$$L = p_{ij}\dot{x}^i\dot{x}^j = R^2\dot{\theta}^2 + R^2\sin^2\theta\dot{\phi}^2,$$

where $i, j = 1, 2$. Applying the Lagrange equations,

$$\frac{\partial L}{\partial x^l} - \frac{d}{d\lambda}\frac{\partial L}{\partial \dot{x}^l} = 0,$$

yields, for each coordinate θ and ϕ,

$$\frac{\partial L}{\partial \theta} - \frac{d}{d\lambda}\frac{\partial L}{\partial \dot{\theta}} = 2R^2\sin\theta\cos\theta\dot{\phi}^2 - 2R^2\ddot{\theta} = 0 \quad \rightarrow \quad \ddot{\theta} = \sin\theta\cos\theta\dot{\phi}^2,$$

$$\frac{\partial L}{\partial \phi} - \frac{d}{d\lambda}\frac{\partial L}{\partial \dot{\phi}} = -4R^2\sin\theta\cos\theta\dot{\theta}\dot{\phi} - 2r^2\sin^2\theta\ddot{\phi} = 0 \quad \rightarrow \quad \ddot{\phi} = -2\cot\theta\dot{\theta}\dot{\phi}.$$

The second equation reduces to

$$\ddot{\phi} - 2\cot\theta\dot{\theta}\dot{\phi} = 0,$$

$$\ddot{\phi}\sin^2\theta - 2\sin\theta\cos\theta\dot{\theta}\dot{\phi} = 0,$$

$$\frac{d}{dt}\left(\dot{\phi}\sin^2\theta\right) = 0,$$

where the conserved term in parentheses is the angular momentum.

We know that the geodesics on the surface of the sphere (shown in Fig. D.1) must be a part of a *great circle* – the circle that contains the two points and whose radius is the radius of the sphere (its center also coincides with the center of the sphere). We can check the two special cases, and make sure they are correct:

1. *Equator:* for the two points along the equator the shortest distance will be also along the equator. We need to show that such a curve $\phi = c_1\lambda + \phi_0$, and $\theta = \pi/2$ satisfies

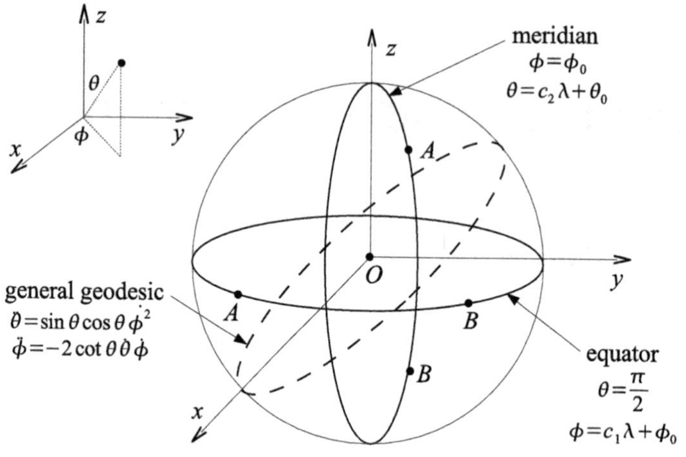

Figure D.1 Geodesic on a sphere.

the geodesic equation. After inserting

$$\begin{aligned} &\phi = c_1\lambda + \phi_0, && \dot{\phi} = c_1, && \ddot{\phi} = 0, \\ &\theta = \tfrac{\pi}{2}, && \dot{\theta} = 0, && \ddot{\theta} = 0, \end{aligned}$$

into the geodesic equation, we obtain

$$\ddot{\theta} = \sin\theta \cos\theta \dot{\phi}^2 = \sin\frac{\pi}{2}\cos\frac{\pi}{2}{c_1}^2 = 0,$$

$$\ddot{\phi} = -2\cot\theta\dot{\theta}\dot{\phi} = -2\cot\frac{\pi}{2}0c_1 = 0.$$

Hence, the equator is a geodesic.

2. *Meridian:* for the two points along the same meridian (arc of the great circle connecting the two poles) the shortest distance should also be along the meridian. We need to show that such a curve $\phi = \phi_0$, and $\theta = c_2\lambda + \theta_0$ satisfies the geodesic equation. After inserting

$$\begin{aligned} &\phi = \phi_0, && \dot{\phi} = 0, && \ddot{\phi} = 0, \\ &\theta = c_2\lambda + \theta_0, && \dot{\theta} = c_2, && \ddot{\theta} = 0, \end{aligned}$$

into the geodesic equation, we obtain

$$\ddot{\theta} = \sin\theta \cos\theta \dot{\phi}^2 = \sin(c_2\lambda + \theta_0)\cos(c_2\lambda + \theta_0)\,0^2 = 0,$$

$$\ddot{\phi} = -2\cot\theta\dot{\theta}\dot{\phi} = -2\cot(c_2\lambda + \theta_0)\,c_2 0 = 0.$$

Hence, the meridian is also a geodesic.

APPENDIX E

Matter–dark energy equality

When did the energy density of matter become equal to the "vacuum" (dark) energy density?

This can be computed easily after recalling that

$$u_\Lambda = const. = u_{\Lambda,0},$$
$$u_\mathrm{m} a^3 = const. \quad \rightarrow \quad u_\mathrm{m} a^3 = u_{\mathrm{m},0} a_0^3 \quad \rightarrow \quad u_\mathrm{m} = u_{\mathrm{m},0} a^{-3},$$

after noting that, by convention, $a_0 = 1$. Therefore the two energy densities are equal at $a_{\mathrm{eq}2}$ when

$$1 = \frac{u_\Lambda}{u_\mathrm{m}} = \frac{u_\Lambda}{u_{\mathrm{m},0} a_{\mathrm{eq}2}^{-3}}$$
$$\rightarrow \quad a_{\mathrm{eq}2} = \left(\frac{u_{\mathrm{m},0}}{u_\Lambda}\right)^{1/3} = \left(\frac{0.28}{0.72}\right)^{1/3} = 0.73.$$

Hence, the energy density of matter and the energy density of dark energy were equal when the Universe was 0.73 – almost 3/4 – of its size today.

To compute how long ago this took place, we can compute the age of the Universe at $a_{\mathrm{eq}2}$ from Eq. (9.23),

$$H_0 t_0 = \int_0^1 \frac{da}{\sqrt{\frac{1-\Omega_\Lambda}{a} + \Omega_\Lambda a^2}} = \int_0^1 \frac{a^{1/2}\, da}{\sqrt{(1-\Omega_\Lambda) + \Omega_\Lambda a^3}}$$
$$= \frac{2}{3\sqrt{\Omega_\Lambda}} \ln\left(2\left(\sqrt{\Omega_\Lambda a^3} + \sqrt{\Omega_\Lambda (a^3 - 1) + 1}\right)\right)\Big|_0^1,$$

by changing the upper limits of integration from t_0 and $a(t_0) = 1$ to t_1 and $a(t_1) \equiv a_{\mathrm{eq}2}$:

$$H_0 t_1 = \int_0^{a_{\mathrm{eq}2}} \frac{da}{\sqrt{\frac{1-\Omega_\Lambda}{a} + \Omega_\Lambda a^2}} = \int_0^{a_{\mathrm{eq}2}} \frac{a^{1/2}\, da}{\sqrt{(1-\Omega_\Lambda) + \Omega_\Lambda a^3}}$$
$$= \frac{2}{3\sqrt{\Omega_\Lambda}} \ln\left(2\left(\sqrt{\Omega_\Lambda a^3} + \sqrt{\Omega_\Lambda (a^3 - 1) + 1}\right)\right)\Big|_0^{a_{\mathrm{eq}2}}$$
$$= \frac{2}{3\sqrt{\Omega_\Lambda}} \ln\left(\frac{\sqrt{\Omega_\Lambda a_{\mathrm{eq}2}^3} + \sqrt{\Omega_\Lambda \left(a_{\mathrm{eq}2}^3 - 1\right) + 1}}{\sqrt{1-\Omega_\Lambda}}\right).$$

Hence, for the observed parameters of $\Omega_\Lambda = 0.72$ and the computed value of $a_{eq2} = 0.73$, we obtain

$$t_1 = \frac{2}{3H_0\sqrt{\Omega_\Lambda}} \ln\left(\frac{\sqrt{0.72\,0.73^3} + \sqrt{0.72\,(0.73^3 - 1) + 1}}{\sqrt{1 - 0.72}}\right) = \frac{2}{3H_0\sqrt{\Omega_\Lambda}}(0.881).$$

We compare this to the age of the Universe computed earlier in Eq. (9.24),

$$t_0 = \frac{2}{3H_0\sqrt{\Omega_\Lambda}} \ln\left(\frac{1 + \sqrt{\Omega_\Lambda}}{\sqrt{1 - \Omega_\Lambda}}\right) = \frac{2}{3H_0\sqrt{\Omega_\Lambda}}(1.25) = 13.7\ \mathcal{A},$$

to finally obtain

$$\frac{t_1}{0.867} = \frac{t_0}{1.25} \qquad \Rightarrow \qquad t_1 = \frac{0.867}{1.25}t_0 = 0.69t_0 = 9.65\ \mathcal{A}.$$

Hence, the Universe was 9.65 billion years old when the energy densities of matter and dark energy were equal. That was $13.7 - 9.65 = 4.05$ billion years ago.

APPENDIX F

Radiation–dark energy equality

When did the energy density of radiation become equal to the "vacuum" (dark) energy density?

The total energy density of radiation is the sum of the energy density of CMB photons, given in Eq. (12.26), and the energy density of neutrinos, given in Eq. (12.41), while the energy density of dark energy is $\Omega_\Lambda = \Omega_{\Lambda,0} = const.$ We then have

$$1 = \frac{\Omega_r}{\Omega_\Lambda} = \frac{\Omega_\gamma + \Omega_\nu}{\Omega_\Lambda} = \frac{\frac{2.48\times10^{-5}}{h^2 a^4} + \frac{1.69\times10^{-5}}{h^2 a^4}}{\Omega_\Lambda} = \frac{4.17\times10^{-5}}{\Omega_\Lambda h^2 a^4}$$

$$\rightarrow \quad a_{eq3} = \left(\frac{4.17\times10^{-5}}{0.72\ 0.73^2}\right)^{1/4} \quad (\approx 0.1)$$

$$\rightarrow \quad 1 + z_{eq3} = \left(\frac{4.17\times10^{-5}}{0.72\ 0.73^2}\right)^{-1/4} \quad (\approx 9.8),$$

where the numbers in parentheses are given for $\Omega_\Lambda = 0.72$ and $h = 0.73$.

From Friedmann's first equation:

$$\left(\frac{\dot{a}}{a}\right)^2 = H_0^2\left(\Omega_{m,0}a^{-3} + \Omega_{r,0}a^{-4} + \Omega_\Lambda\right).$$

Solving for $\dot{a}$, this becomes

$$\dot{a} = H_0\sqrt{\frac{\Omega_{m,0}}{a} + \frac{\Omega_{r,0}}{a^2} + \Omega_\Lambda a^2}$$

and

$$H_0 t_{eq3} = \int_0^{a_{eq3}} \frac{da}{\sqrt{\frac{\Omega_{m,0}}{a} + \frac{\Omega_{r,0}}{a^2} + \Omega_\Lambda a^2}} = \int_0^{a_{eq3}} \frac{a\ da}{\sqrt{\Omega_{m,0}a + \Omega_{r,0} + \Omega_\Lambda a^4}},$$

hence, the age of the Universe for $\Omega_\Lambda = 0.72$, $\Omega_{m,0} = 0.28$, and $\Omega_{r,0} = 4.17\times10^{-5}/h^2 = 7.8\times10^{-5}$ at a_{eq3} is (after using your favorite computational tool, such as Mathematica, Matlab®, or Python, to perform the calculation):

$$t_{eq3} \approx 0.548\ \mathcal{A} \approx 5.48\times10^8\ \text{yr} = 548\ \text{million years}.$$

APPENDIX G

Radiation–matter equality

It is beneficial to compute at which point the energy densities of radiation and matter were equal, because that was the point of transition between these two different regimes. This point is called *radiation–matter equality*. The significance of this transition is that the perturbations in the two regimes grow at different rates, as we will see later.

We find the value of the scale factor $a(t) = a_{\text{eq}}$ at which the energy densities of matter and radiation were equal by setting their ratio to unity and solving for a.

The total energy density of radiation is the sum of the energy density of CMB photons, given in Eq. (12.26), and the energy density of neutrinos, given in Eq. (12.41), while the energy density of baryons is given in Eq. (12.45). We then have

$$
\begin{aligned}
1 = \frac{\Omega_\gamma + \Omega_\nu}{\Omega_{\text{m}}} &= \frac{\frac{2.48\times10^{-5}}{h^2a^4} + \frac{1.69\times10^{-5}}{h^2a^4}}{\Omega_{\text{m},0}a^{-3}} = \frac{4.17\times10^{-5}}{\Omega_{\text{m},0}h^2a} \\
\rightarrow \quad a_{\text{eq}} &= \frac{4.17\times10^{-5}}{\Omega_{\text{m},0}h^2}\left(=2.79\times10^{-4}\right) \\
\rightarrow \quad 1+z_{\text{eq}} &= 2.43\times10^{4}\Omega_{\text{m},0}h^2\left(=3.58\times10^{3}\right),
\end{aligned}
$$

where the numbers in parentheses are given for $\Omega_{\text{m},0} = 0.28$ and $h = 0.73$. We saw earlier that the photons decouple from matter around $z \approx 10^3$, *after* the matter–radiation equality, which means that the decoupling takes place in a matter-dominated Universe.

Let us now estimate how old the Universe was when this happened. From Friedmann's first equation:

$$
\left(\frac{\dot{a}}{a}\right)^2 = H_0^2\left[\Omega_{\text{m},0}a^{-3} + \Omega_{\text{r},0}a^{-4} + \Omega_\Lambda\right].
$$

Solving for $\dot{a}$, this becomes

$$
\dot{a} = H_0\sqrt{\frac{\Omega_{\text{m},0}}{a} + \frac{\Omega_{\text{r},0}}{a^2} + \Omega_\Lambda a^2}
$$

and

$$
H_0 t_{\text{eq}} = \int_0^{a_{\text{eq3}}} \frac{da}{\sqrt{\frac{\Omega_{\text{m},0}}{a} + \frac{\Omega_{\text{r},0}}{a^2} + \Omega_\Lambda a^2}} = \int_0^{a_{\text{eq3}}} \frac{a\,da}{\sqrt{\Omega_{\text{m},0}a + \Omega_{\text{r},0} + \Omega_\Lambda a^4}},
$$

hence, the age of the Universe for $\Omega_\Lambda = 0.72$, $\Omega_{\text{m},0} = 0.28$, and $\Omega_{\text{r},0} = 4.17\times10^{-5}/h^2 = 7.8\times10^{-5}$ at a_{eq} is (after using your favorite computational tool, such as Mathematica,

Matlab®, or Python, to perform the calculation):

$$t_0 \approx 4.62 \times 10^{-5} \mathcal{A} = 4.62 \times 10^4 \text{ yr} \approx 46000 \text{ yr}.$$

APPENDIX H

Chemical potential

The distribution function for species for both fermions and bosons is given by

$$f = \frac{1}{e^{(E-\mu)/T} \pm 1},$$

(+ for fermions and − for bosons). For a thermal background radiation, the *chemical potential* μ is always zero. The reason for this is the following: μ is defined in the context of the first law of thermodynamics as the change in energy associated with the change in particle number

$$dE = TdS - PdV + \mu dN.$$

As N adjusts to its equilibrium value, we expect that the system will be stationary with respect to small changes in N. More rigorously, the Helmholtz free energy $F = E - TS$ is minimized ($dF/dN = 0$) in equilibrium for a system at constant temperature ($dT = 0$) and volume ($dV = 0$). Taking the derivative of the Helmholtz energy, we obtain

$$dF = dE - TdS - SdT,$$

which, combined with the equation for dE above, yields

$$dF = TdS - PdV + \mu dN - TdS - SdT = -PdV - SdT + \mu dN$$
$$\rightarrow \quad \frac{dF}{dN} = -P\frac{dV}{dN} - S\frac{dT}{dN} + \mu = \mu = 0.$$

APPENDIX I

How to compute the relative abundances of the light elements

Step 1: Calculate the abundance ratio n_n/n_p (ratio of the number of neutrons n_n to the number of protons n_p) as a function of time by considering the reaction $n \leftrightarrow p + e^- + \bar{\nu}_e$. This entails considering all the possible permutations, namely

$$n + \nu_e \to p + e^-, \qquad n + e^+ \to p + \bar{\nu}_e, \qquad n \to p + e^- + \bar{\nu}_e, \qquad \text{and}$$
$$p + e^- \to n + \nu_e, \qquad p + \bar{\nu}_e \to n + e^+, \qquad p + e^- + \bar{\nu}_e \to n,$$

which leads to a coupled system of rate equations that involve (1) the particle energies, (2) the particle densities, and (3) the assumed interaction mechanisms. One assumes typically that the Universe is completely homogeneous and isotropic. The energies follow from the assumption that everything is in thermal equilibrium, so that the distribution of energies is a function of temperature T. The particle densities vary because of the expansion of the Universe, satisfying $n \propto a^{-3} \propto T^3$. To sufficient accuracy, the interaction mechanisms are described by the old-fashioned $V - A$ theory of the weak interaction dating back to the 1970s. The equations are solved subject to an initial condition that

$$\frac{n_n}{n_p}(t_0) = \frac{\exp(-m_n c^2/kT)}{\exp(-m_p c^2/kT)} = \exp(-\Delta mc^2/kT),$$

at some temperature $kT \gg \Delta mc^2$.

For sufficiently high temperature T, one finds that the density remains in equilibrium, but when the temperature and density have become too low, the system falls out of equilibrium and the ratio becomes frozen at a value

$$\frac{n_n}{n_p} = \exp(-\Delta mc^2/kT_*),$$

where $T_* \approx 0.9 \times 10^{10}$ K. After this happens, the only relevant point is that, unless a neutron has become trapped in the nucleus, it will decay with a half-life of about 10.6 min. (Perhaps surprisingly, the exact value of this half-life is not all that well known, since it is difficult to measure experimentally!)

Step 2: Calculate the elemental abundances, assuming that n_n/n_p is known as a function of T or t. To do this correctly, one needs to write down and solve all the possible reaction equations of nuclear physics, which entails a huge coupled system of ordinary differential equations that translate all dependence on energy and density into a dependence on temperature T. However, as a practical matter, it turns out that only a few of the equations are actually important.

The net result of such a computation is that almost all the neutrons that do not decay end up in ^{4}He nuclei: it is easy to convert neutrons and protons to ^{4}He. The first stage is to convert protons and neutrons into deuterium (d), and then convert deuterium into tritium (^{3}H) or helium-3 (^{3}He):

$$p + n \rightarrow d, \qquad d + n \rightarrow {}^3\mathrm{H}, \qquad 2d \rightarrow {}^3\mathrm{H} + p, \qquad \text{and} \qquad 2d \rightarrow {}^3\mathrm{He} + n.$$

What remains then is to convert ^{3}H and ^{3}He into ^{4}He through the interactions

$$^3\mathrm{H} + p \rightarrow {}^4\mathrm{He}, \qquad \text{or} \qquad {}^3\mathrm{H} + d \rightarrow {}^4\mathrm{He} + n, \qquad \text{and}$$
$$^3\mathrm{He} + n \rightarrow {}^4\mathrm{He}, \qquad {}^3\mathrm{He} + d \rightarrow {}^4\mathrm{He} + p, \qquad \text{or} \qquad {}^3\mathrm{He} + {}^3\mathrm{He} \rightarrow {}^4\mathrm{He} + 2p.$$

Since ^{4}He is a very stable nucleus, it will not decay back into smaller constituents. However, combining ^{4}He with anything else to make heavier nuclei is very difficult, since all the possible nuclei containing $A = 5$ or $A = 8$ nucleons are extremely unstable. Essentially, nothing can be made via the reaction

$$^4\mathrm{H} + p \rightarrow, \qquad {}^4\mathrm{H} + n \rightarrow, \qquad \text{or} \qquad {}^4\mathrm{He} + {}^4\mathrm{He} \rightarrow .$$

(In aging stars, helium fuses via a triple-alpha process that involves combining three ^{4}He nuclei.) One *does* manufacture tiny amounts of ^{7}Li via the interactions

$$^4\mathrm{He} + {}^3\mathrm{H} \rightarrow {}^7\mathrm{Li}, \qquad {}^4\mathrm{He} + {}^3\mathrm{He} \rightarrow {}^7\mathrm{Be}, \qquad {}^7\mathrm{Be} + e^- \rightarrow {}^7\mathrm{Li},$$

however, the relative abundance thereof is only $\sim 10^{-10} - 10^{-9}$ as mass fraction.

Dependence on various inputs

The Role of Photons:

1. The photons, neutrinos, and anti-neutrinos are the dominant contribution to the quantities ρ and P, which, in turn, determine $a(t)$, the rate at which the Universe expands.
2. At high temperatures, deuterium nuclei that are created initially can be split back into a proton and a neutron by a photon with energy $E \gtrsim 2.2$ MeV. If the numbers of photons and nucleons were comparable, this would mean that deuterium could start to remain stuck together at $kT \lesssim 2$ MeV. Given, however, that there are literally billions of photons for each nucleon, such sticking requires a considerably lower temperature, $kT \lesssim 0.1$ MeV. The point is that the number of photons per baryon

with energies in excess of 2.2 MeV scales as

$$\frac{n_\gamma}{n_B}\int_{2.2\ \mathrm{MeV}}^{\infty} f(E)dE \approx \frac{n_\gamma}{n_B}\exp(2.2\ \mathrm{MeV}/kT),$$

where n_γ is the number of photons, and n_B the number of baryons. The ratio n_γ/n_B is thus crucial in determining when deuterium can survive long enough to combine into another, larger nucleus. If the value of this ratio were to increase, so that there were more photons per baryon, it would be harder for deuterium to remain stuck together.

The Half-Life of the Neutron:
Free neutrons are unstable toward $n \to p + e + \bar{\nu}_e$, with a half-life of about $\tau_n \approx 10.6$ min. If neutrons were not combined into nuclei quickly enough, they would all decay into protons and there would be no heavy elements created primordially. All else being equal, smaller τ_n would mean fewer heavier elements.

The Number of Neutrino Flavors:
N_ν fixes the total energy density ρ, and hence determines the overall expansion rate.

Basic trends

Deuterium and ^{3}He are transformed into ^{4}He by capturing a proton or neutron. A higher density of baryons means that there are more nearby protons and neutrons, so that conversion happens more quickly and less deuterium and ^{3}He will remain. In other words,

$$\frac{dn_{\mathrm{D}}}{dn_B} < 0, \qquad \text{and} \qquad \frac{dn_{^3\mathrm{He}}}{dn_B} < 0.$$

When n_B/n_γ is small, lithium is created and destroyed primarily via the reactions

$$^4\mathrm{He} + ^3\mathrm{H} \to ^7\mathrm{Li}, \qquad \text{and} \qquad ^7\mathrm{Li} + p \to 2\ ^4\mathrm{He}.$$

It follows that an *increase in n_B/n_γ causes a decrease in the abundance of lithium* because there are more protons to destroy ^{7}Li.

Alternatively, for larger n_B/n_γ, lithium is manufactured primarily via the reactions

$$^4\mathrm{He} + ^3\mathrm{He} \to ^7\mathrm{Be}, \qquad \text{and} \qquad ^7\mathrm{Be} + e^- \to ^7\mathrm{Li}.$$

In this case, *increasing n_B/n_γ causes an increase in the abundance of lithium, since there are more nucleons to make* ^{3}He. The net result is that a graph of the relative abundance of ^{7}Li as a function of n_B exhibits a sharp dip at a value corresponding to $\Omega_B h^2 \approx 0.01$, where Ω_B is the mass density $\rho_B = m_p n_B$ of baryons expressed in units of critical density.

Almost all the neutrinos that did not decay are ultimately converted into ^{4}He, so that the abundance $n_{^4\text{He}}$ is controlled largely by how many neutrons survive. If N_ν were larger, ρ would be larger, so that $\dot{a}/a$ would be larger and the Universe would expand more quickly. In this case, neutrons and protons would fall out of equilibrium earlier on, when the ratio n_n/n_p was larger, so that there would be more neutrons overall. Hence, the relative abundance of ^{4}He should increase. N_ν also plays a role in the abundance of d and ^{3}He, since it provides an interplay between the expansion rate and the reaction rate that fixes the abundances of those elements.

The obvious question is: *Is there a seemingly realistic choice of parameters that would lead to predicted elemental abundances in agreement with the elemental abundances observed and/or inferred in the least processed parts of the Universe that we can detect?* The answer here is: yes! Assuming that $N_\nu = 3$ and $\tau_n = 10.6$ min, one can explain the best observational estimates, namely

$$0.24 < Y_{^4\text{He}} < 0.26, \qquad \text{and} \qquad 1 < \frac{n_{d+^3\text{He}}}{n_p} < 20,$$

by determining that

$$n_B = 1.4 \pm 0.3 \times 10^7 \text{ nucleons/cm}^3 \qquad \text{and} \qquad \Omega_B = \frac{m_p n_B}{\rho_c} = 0.013 \pm 0.003\ h^{-2}.$$

This corresponds to a mass density considerably larger than the amount of luminous matter detected directly in individual galaxies. However, it is substantially smaller than the density inferred from dynamical estimates of the masses of galaxies based on an analysis of the motions of luminous objects in galaxies.

Realistic uncertainties

- Perhaps surprisingly, the reaction rate $^3\text{He} + d \rightarrow {}^4\text{He} + n$ is only known to within a factor of 2–3, which leads to substantial uncertainties in the abundance of ^{7}Li.
- Again perhaps surprisingly, the half-life of the neutron is only known to ± 0.5 min; however, for example, a half-life as short as $\tau_n = 10.2$ min would yield a change $\Delta Y_{^4\text{He}} \approx 0.006$.
- If one were to discover a new species of neutrino, or any other particle with $mc^2 \ll kT$ for $T \approx 10^{11}$ K, the predicted abundances would change significantly.
- If the Universe were significantly anisotropic or inhomogeneous early on, the abundances of the light elements would be different at different places in the Universe and, even if things were later homogenized, the average rates would differ from those predicted assuming a homogeneous and isotropic Universe.

APPENDIX J

Equation of state for the perfect fluid

In order to solve the set of Friedmann's equations for the scale factor of the Universe a, we must complete the set with an equation of state $P(\rho)$ that defines the pressure provided by the distribution of matter used in the model. For the dust model, in which matter sits still in its comoving reference frame $u_\alpha = (c, 0, 0, 0)$, there is no pressure that matter exerts: $P = 0$. However, the perfect fluid of radiation particles (photons and neutrinos) does exert non-zero pressure, and in this appendix we derive the corresponding equation of state from first principles.

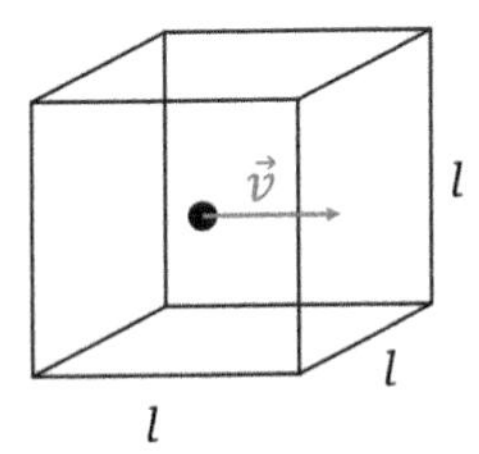

Figure J.1 Pressure exerted by a relativistic particle moving at the speed v on a fiducial box with sides of length l.

In order to compute the radiation pressure due to an ideal gas of photons, let us first define the *ideal gas* as a collection of N identical relativistic $v = c$ particles, which at every point satisfy the ideal-gas equation,

$$PV = Nkt.$$

Let us compute a pressure on the walls of a fiducial box with sides of length l shown in Fig. J.1, and then average it over the entire distribution of N randomly distributed particles.

The particle travels from one side of the box to the other (say, from left to right) in

$$\Delta t = \frac{l}{c}.$$

As the particle with momentum $p = E/c$ collides with the wall on the right, it exerts a force on it,

$$F = \frac{\Delta p}{\Delta t} = \frac{2p}{l/c} = \frac{2pc}{l}.$$

The pressure, defined as force per unit area A, is then

$$P = \frac{F}{A} = \frac{2pc}{l^3} = \frac{2E}{l^3}.$$

However, only 1/6 of all the randomly distributed particles are going in to hit the right-most wall, hence,

$$P = \frac{1}{6}\frac{2E}{l^3} = \frac{1}{3}\frac{E}{V} = \frac{1}{3}u.$$

APPENDIX K

Origins of the large-scale structure in the Universe

The gravitational instability picture

Although the Universe appears to be almost completely homogeneous and isotropic on the largest scales, it is clear that it is decidedly lumpy on smaller scales, extending from galaxy clusters down to the scales of stars and planets. The obvious question, therefore, is how is one to describe, and account for, these observed structures in the context of a Universe that is reasonably well approximated by a Friedmann cosmology on very large scales.

The currently accepted paradigm is known as the *gravitational instability* picture. The idea is that, very early on, the Universe was almost, but not exactly, homogeneous and isotropic, and that the small irregularities present early on were amplified by virtue of the fact that a nearly homogeneous and isotropic Universe is unstable dynamically.

Equations of structure in a Newtonian limit

According to the gravitational instability picture, a correct description of the origins of large-scale structure would involve formulating all the equations of general relativity for a system that is almost, but not exactly, homogeneous and isotropic. Not surprisingly, this quickly becomes quite messy, since one has relaxed the assumed symmetries that make the Friedmann cosmologies comparatively simple. Fortunately, however, such a tack is not necessary. If one restricts attention to scales small compared with the horizon length, *i.e.*, $r \ll ct_H$, where t_H is the age of the Universe at the time in question and, in addition, assumes that the matter that is to be described is characterized by peculiar motions small compared with the speed of light, then the requisite equations can be derived in the context of Newtonian swindle whereby the Universe is viewed as a gigantic explosion.

The Newtonian pictures visualize the Universe as expanding relative to some fictitious inertial frame, so that a collection of objects with no peculiar velocities, *i.e.*, no motions superimposed on the overall expansion, will preserve their relative orientation as the explosion progresses. It is thus useful to reformulate the Newtonian model in terms of comoving coordinates by writing the physical distance $\boldsymbol{r}$ in the form $\boldsymbol{r} = a(t)\boldsymbol{x}$. This implies that the physical velocity $\boldsymbol{u}$ satisfies

$$\boldsymbol{u} = a\dot{\boldsymbol{x}} + \boldsymbol{x}\dot{a},$$

a relation that makes explicit the fact that the motion of any object entails both a peculiar velocity $a\dot{\boldsymbol{x}}$ and the systemic Hubble velocity $\boldsymbol{x}\dot{a}$.

As formulated in comoving coordinates, the equation of motion for a particle moving in a gravitational potential Φ can be derived from a Lagrangian of the form

$$L = \frac{1}{2}m\boldsymbol{u}^2 - m\Phi(\boldsymbol{x}, t) = \frac{1}{2}m\left(a\dot{\boldsymbol{x}} + \boldsymbol{x}\dot{a}\right)^2 - m\Phi(\boldsymbol{x}, t).$$

Applying the Lagrange equation,

$$\frac{\partial L}{\partial \boldsymbol{x}} - \frac{d}{dt}\frac{\partial L}{\partial \dot{\boldsymbol{x}}} = 0,$$

leads to

$$ma\left(a\ddot{\boldsymbol{x}} + 2\dot{a}\dot{\boldsymbol{x}} + \ddot{a}\boldsymbol{x}\right) = -m\frac{\partial \Phi}{\partial \boldsymbol{x}}.$$

Recalling, however, that the peculiar velocity $\boldsymbol{v} = a\dot{\boldsymbol{x}}$, it is clear that

$$\frac{d\boldsymbol{v}}{dt} = \dot{a}\dot{\boldsymbol{x}} + a\ddot{\boldsymbol{x}},$$

so that the equation of motion can be written as

$$ma\left(\frac{d\boldsymbol{v}}{dt} + \boldsymbol{v}\frac{\dot{a}}{a}\right) = -m\frac{\partial \Phi}{\partial \boldsymbol{x}} - ma\ddot{a}\boldsymbol{x} \equiv -m\frac{\partial \Psi}{\partial \boldsymbol{x}},$$

where

$$\Psi(\boldsymbol{x}, t) \equiv \Phi(\boldsymbol{x}, t) + \frac{1}{2}a\ddot{a}\boldsymbol{x}^2.$$

Given this equation for the motion of a single particle, it should be clear that a collection of particles idealized as a perfect fluid should satisfy an Euler equation of the form,

$$\rho a\left(\frac{d\boldsymbol{v}}{dt} + \frac{\dot{a}}{a}\boldsymbol{v}\right) = -\rho\nabla\Psi - \nabla P,$$

where $\nabla \equiv \partial/\partial\boldsymbol{x}$. Alternatively, by expanding out the total time derivative, one has

$$\frac{\partial \boldsymbol{v}}{\partial t} + \frac{\dot{a}}{a}\boldsymbol{v} + \frac{1}{a}\left(\boldsymbol{v}\nabla\right)\boldsymbol{v} = -\frac{1}{a}\nabla\Psi - \frac{1}{a\rho}\nabla P.$$

Here, the term "quadratic" in $\boldsymbol{v}$ comes with a factor of $1/a$ because $d\boldsymbol{x}/dt = \boldsymbol{v}/a$.

The form of the Poisson equation in comoving coordinates is also easily derived. Expressed in physical variables, the Poisson equation takes the form

$$\nabla_r^2\Phi = 4\pi G\rho,$$

where $\nabla_r \equiv \partial/\partial \boldsymbol{r}$. Rewriting the left-hand side in terms of ∇ yields a factor of $1/a^2$, so that

$$\nabla^2\Phi = 4\pi G\rho a^2 \quad \rightarrow \quad \nabla^2\Psi = 4\pi G\rho a^2 + 3a\ddot{a}.$$

Given, however, that the scale factor satisfies the relation

$$\ddot{a} = -\frac{4\pi G\rho_0}{3}a,$$

it follows that

$$\nabla^2\Psi = 4\pi G\rho a^2\left[\rho(\boldsymbol{x},t) - \rho_0(t)\right],$$

where $\rho_0(t)$ is the time-dependent average density of the expanding system. It follows that Ψ can be interpreted as the gravitational potential associated with any irregularities in the mass distribution. Expressed in physical coordinates, the equation expressing conservation of mass reads

$$\frac{\partial\rho}{\partial t_r} + \nabla_r(\rho\boldsymbol{u}) = 0,$$

where the time derivative is to be performed for fixed $\boldsymbol{r}$. Given, however, that one wants to view ρ as a function of $\boldsymbol{x} = \boldsymbol{r}/a$, it is clear that

$$\left(\frac{\partial}{\partial t}\right)_r \rho(\boldsymbol{r}/a(t),t) = \frac{\partial\rho}{\partial t} - \frac{\dot{a}}{a}\boldsymbol{x}\nabla\rho$$

and, similarly, since $\boldsymbol{u} = \boldsymbol{v} + a\boldsymbol{x}$,

$$\nabla_r(\rho\boldsymbol{u}) = \frac{1}{a}\nabla(\rho\boldsymbol{v}) + 3\frac{\dot{a}}{a}\rho + \frac{\dot{a}}{a}\boldsymbol{x}\nabla\rho.$$

Combining these relations leads immediately to an equation of the form

$$\frac{\partial\rho}{\partial t} + +3\frac{\dot{a}}{a}\rho + \frac{1}{a}\nabla(\rho\boldsymbol{v}).$$

Alternatively, in terms of the fractional density, irregularity δ, defined by the relation

$$\rho = \rho_0(1+\delta),$$

the equation of conservation of mass takes the form

$$\frac{\partial\delta}{\partial t} + \frac{1}{a}\nabla\left[(1+\delta)\boldsymbol{v}\right] = 0.$$

The equations derived up until now are exact, at least within the context of the Newtonian model. Unfortunately, however, they are difficult, if not impossible, to solve analytically except for special uninteresting cases such $\boldsymbol{v}=0$, $\delta=0$, $\Psi=0$. One is forced immediately to resort to numerical computations. Given, however, that one is interested in initial conditions corresponding to small deviations from homogeneity and isotropy, one *can* proceed analytically by implementing the approximation that $\boldsymbol{v}$, δ, and Ψ are all small, and that any term in an equation that involves the product of two such small terms can be treated as negligible. In this approximation, the relevant equations reduce to

$$\frac{\partial \delta}{\partial t}+\frac{1}{a}\nabla\left[(1+\delta)\boldsymbol{v}\right]=0.$$

The equations derived up until now are exact, at least within the

$$\frac{\partial \boldsymbol{v}}{\partial t}+\frac{\dot{a}}{a}\boldsymbol{v}=-\frac{1}{a}\nabla\Psi-\frac{1}{a\rho_0}\nabla\delta P, \tag{K.1}$$

$$\nabla^2\Psi=4\pi Ga^2\rho_0\delta$$

and

$$\frac{\partial \delta}{\partial t}+\frac{1}{a}\nabla\boldsymbol{v}=0,$$

where δP is the irregularity associated with the density irregularity δ, *i.e.*,

$$\rho=\rho_0(t)(1+\delta(\boldsymbol{x},t)) \quad \rightarrow \quad P=P_0(t)+\delta P(\boldsymbol{x},t).$$

The Jeans swindle

These equations are especially simple to solve if one assumes, unrealistically, that $a(t)$ is actually a time-independent constant. This is, of course, logically inconsistent, since it was assumed in the formulation of these equations that $\ddot{a}=4\pi G\rho_0 a/3$. Given this assumption, one can always choose units in which $a\equiv 1$, so that the equations reduced to

$$\frac{\partial \boldsymbol{v}}{\partial t}=-\nabla\Psi-1\rho_0\nabla\delta P,$$

$$\nabla^2\Psi=4\pi Ga^2\rho_0\delta,$$

$$\frac{\partial \delta\rho}{\partial t}+\rho_0\nabla\boldsymbol{v}=0,$$

where, for historical reasons, one has written $\delta\rho=\rho\delta$. Differentiating the equation expressing conservation of mass with respect to time t leads to the relation

$$\frac{\partial^2\delta\rho}{\partial t^2}+\rho_0\nabla\frac{\partial \boldsymbol{v}}{\partial t}=\frac{\partial^2\delta\rho}{\partial t^2}+\rho_0\nabla\left(-\frac{1}{\rho_0}\nabla\delta P-\nabla\Psi\right)=0.$$

APPENDIX K

Origins of the large-scale structure in the Universe

The gravitational instability picture

Although the Universe appears to be almost completely homogeneous and isotropic on the largest scales, it is clear that it is decidedly lumpy on smaller scales, extending from galaxy clusters down to the scales of stars and planets. The obvious question, therefore, is how is one to describe, and account for, these observed structures in the context of a Universe that is reasonably well approximated by a Friedmann cosmology on very large scales.

The currently accepted paradigm is known as the *gravitational instability* picture. The idea is that, very early on, the Universe was almost, but not exactly, homogeneous and isotropic, and that the small irregularities present early on were amplified by virtue of the fact that a nearly homogeneous and isotropic Universe is unstable dynamically.

Equations of structure in a Newtonian limit

According to the gravitational instability picture, a correct description of the origins of large-scale structure would involve formulating all the equations of general relativity for a system that is almost, but not exactly, homogeneous and isotropic. Not surprisingly, this quickly becomes quite messy, since one has relaxed the assumed symmetries that make the Friedmann cosmologies comparatively simple. Fortunately, however, such a tack is not necessary. If one restricts attention to scales small compared with the horizon length, *i.e.*, $r \ll ct_{\rm H}$, where $t_{\rm H}$ is the age of the Universe at the time in question and, in addition, assumes that the matter that is to be described is characterized by peculiar motions small compared with the speed of light, then the requisite equations can be derived in the context of Newtonian swindle whereby the Universe is viewed as a gigantic explosion.

The Newtonian pictures visualize the Universe as expanding relative to some fictitious inertial frame, so that a collection of objects with no peculiar velocities, *i.e.*, no motions superimposed on the overall expansion, will preserve their relative orientation as the explosion progresses. It is thus useful to reformulate the Newtonian model in terms of comoving coordinates by writing the physical distance $\boldsymbol{r}$ in the form $\boldsymbol{r} = a(t)\boldsymbol{x}$. This implies that the physical velocity $\boldsymbol{u}$ satisfies

$$\boldsymbol{u} = a\dot{\boldsymbol{x}} + \boldsymbol{x}\dot{a},$$

a relation that makes explicit the fact that the motion of any object entails both a peculiar velocity $a\dot{\boldsymbol{x}}$ and the systemic Hubble velocity $\boldsymbol{x}\dot{a}$.

As formulated in comoving coordinates, the equation of motion for a particle moving in a gravitational potential Φ can be derived from a Lagrangian of the form

$$L = \frac{1}{2}m\boldsymbol{u}^2 - m\Phi(\boldsymbol{x}, t) = \frac{1}{2}m\,(a\dot{\boldsymbol{x}} + \boldsymbol{x}\dot{a})^2 - m\Phi(\boldsymbol{x}, t).$$

Applying the Lagrange equation,

$$\frac{\partial L}{\partial \boldsymbol{x}} - \frac{d}{dt}\frac{\partial L}{\partial \dot{\boldsymbol{x}}} = 0,$$

leads to

$$ma\,(a\ddot{\boldsymbol{x}} + 2\dot{a}\dot{\boldsymbol{x}} + \ddot{a}\boldsymbol{x}) = -m\frac{\partial \Phi}{\partial \boldsymbol{x}}.$$

Recalling, however, that the peculiar velocity $\boldsymbol{v} = a\dot{\boldsymbol{x}}$, it is clear that

$$\frac{d\boldsymbol{v}}{dt} = \dot{a}\dot{\boldsymbol{x}} + a\ddot{\boldsymbol{x}},$$

so that the equation of motion can be written as

$$ma\left(\frac{d\boldsymbol{v}}{dt} + \boldsymbol{v}\frac{\dot{a}}{a}\right) = -m\frac{\partial \Phi}{\partial \boldsymbol{x}} - ma\ddot{a}\boldsymbol{x} \equiv -m\frac{\partial \Psi}{\partial \boldsymbol{x}},$$

where

$$\Psi(\boldsymbol{x}, t) \equiv \Phi(\boldsymbol{x}, t) + \frac{1}{2}a\ddot{a}\boldsymbol{x}^2.$$

Given this equation for the motion of a single particle, it should be clear that a collection of particles idealized as a perfect fluid should satisfy an Euler equation of the form,

$$\rho a\left(\frac{d\boldsymbol{v}}{dt} + \frac{\dot{a}}{a}\boldsymbol{v}\right) = -\rho\boldsymbol{\nabla}\Psi - \boldsymbol{\nabla}P,$$

where $\boldsymbol{\nabla} \equiv \partial/\partial\boldsymbol{x}$. Alternatively, by expanding out the total time derivative, one has

$$\frac{\partial \boldsymbol{v}}{\partial t} + \frac{\dot{a}}{a}\boldsymbol{v} + \frac{1}{a}(\boldsymbol{v}\boldsymbol{\nabla})\,\boldsymbol{v} = -\frac{1}{a}\boldsymbol{\nabla}\Psi - \frac{1}{a\rho}\boldsymbol{\nabla}P.$$

Here, the term "quadratic" in $\boldsymbol{v}$ comes with a factor of $1/a$ because $d\boldsymbol{x}/dt = \boldsymbol{v}/a$.

The form of the Poisson equation in comoving coordinates is also easily derived. Expressed in physical variables, the Poisson equation takes the form

$$\boldsymbol{\nabla}_r^2\Phi = 4\pi G\rho,$$

Since, by assumption, ρ_0 is a time-independent constant, so that

$$\frac{\partial^2 \delta\rho}{\partial t^2} - \nabla^2 \delta P = \rho_0 \nabla^2 \Psi = 4\pi G \rho_0 \delta\rho.$$

To solve this equation, one needs to specify an equation of state, $P = P(\rho)$, relating P and ρ. Given such an equation of state, one can write

$$\delta P = \left(\frac{\partial P}{\partial \rho}\right) \delta\rho \equiv c_s^2 \delta\rho,$$

where c_s is the speed of propagation of density waves (*i.e.*, speed of sound). It follows that

$$\frac{\partial^2 \delta\rho}{\partial t^2} - c_s^2 \nabla^2 \delta\rho = 4\pi G \rho_0 \delta\rho.$$

This is a linear partial differential equation for $\delta\rho$ with coefficients that are independent of $\boldsymbol{x}$. For this reason, it is convenient to look for solutions of the form

$$\delta\rho(\boldsymbol{x}, t) = \delta\rho(t) \exp(i\boldsymbol{k}\boldsymbol{x}),$$

where $\boldsymbol{k}$ is the wavenumber. One knows that such plane waves constitute a complete set of orthogonal functions, so that any $\delta\rho$ can be viewed as a superposition of plane waves. Given this form of solutions, the equation for $\delta\rho(t)$ reduces to

$$\frac{d^2 \delta\rho}{dt^2} + c_s^2 k^2 \delta\rho = 4\pi G \rho_0 \delta\rho,$$

or

$$\frac{d^2 \delta\rho}{dt^2} + \omega_k^2 \delta\rho = 0, \quad \text{with} \quad \omega_k^2 \equiv c_s^2 k^2 - 4\pi G \rho_0.$$

For $k^2 \to \infty$, the solutions to this equation correspond to sinusoidal oscillations with a frequency $\omega_k = c_s k$, *i.e.*, the acoustical (sound) waves mentioned earlier. However, for longer wavelengths, corresponding to wavenumbers satisfying

$$k^2 < \frac{4\pi G \rho_0}{c_s^2} \equiv k_J^2,$$

ω_k becomes negative. In this case, the solutions satisfy

$$\delta\rho \propto \exp(\pm \omega_k t),$$

which corresponds to solutions that grow and damp out exponentially in time. The existence of an exponentially growing solution implies that the homogeneous and isotropic solution is unstable. In particular, for the case of very strong wavelength perturbation, $k \to 0$, one has

$$\delta\rho \propto \exp(\pm t/\tau), \quad \text{with} \quad \tau \equiv \frac{1}{\sqrt{4\pi G\rho_0}}.$$

This is the so-called *Jeans instability*, and the critical length

$$R_J = \frac{2\pi}{k_J},$$

is known as the *Jeans length*. The quantity

$$M_J = \frac{4\pi}{3}\rho_0 R_J^3,$$

is the *Jeans mass*.

Constants, units, and conversions

Fundamental constants	
speed of light	$c = 2.998 \times 10^8$ m/s
Newton's constant	$G = 6.673 \times 10^{-11}$ m^3 kg^{-1} s^{-2}
Planck constant	$h = 6.626 \times 10^{-34}$ J s $= 4.136 \times 10^{-15}$ eV s
reduced Planck constant	$\hbar = h/(2\pi) = 1.055 \times 10^{-34}$ J s $= 6.582 \times 10^{-16}$ eV s
Boltzmann constant	$k_B = 1.381 \times 10^{-23}$ m^2 kg s^{-2} K^{-1}
electron mass	$m_e = 9.110 \times 10^{-31}$ kg $= 511$ keV$/c^2$
proton mass	$m_p = 1.673 \times 10^{-27}$ kg $= 938.3$ MeV$/c^2$
neutron mass	$m_n = 1.675 \times 10^{-27}$ kg $= 939.6$ MeV$/c^2$
Sun's mass	$M_\odot = 1.988 \times 10^{30}$ kg
elementary charge	$q = 1.602 \times 10^{-19}$ C
Cosmological parameters	
Hubble parameter (today)	$H_0 \equiv 100\, h$ km/s/Mpc
Hubble time	$H_0^{-1} = 9.78 \times 10^9\, h^{-1}$ yr
Hubble distance	$cH_0^{-1} = 9.78 \times 10^9\, h^{-1}$ ly $= 3.00 \times 10^9\, h^{-1}$ pc
critical mass density (today)	$\rho_{cr,0} = 1.87 \times 10^{-26}\, h^2$ kg/m^3
temperature of the Universe (today)	$T_0 = 2.726$ K
Conversions	
year	1 yr $= 3.154 \times 10^7$ s
light year	1 ly $= 0.3067$ pc $= 9.461 \times 10^{15}$ m
parsec	1 pc $= 3.261$ ly $= 3.086 \times 10^{16}$ m
eV	1 eV $= 1.602 \times 10^{-19}$ J
Planck Units	
Planck time	$t_P = (G\hbar/c^5)^{1/2} = 5.391 \times 10^{-44}$ s
Planck mass	$m_P = (\hbar c/G)^{1/2} = 2.177 \times 10^{-8}$ kg
Planck length	$l_P = (G\hbar/c^3)^{1/2} = 1.616 \times 10^{-35}$ m
Planck energy	$E_P = (\hbar c^5/G)^{1/2} = 1.957 \times 10^9$ J $= 1.221 \times 10^{28}$ eV
Planck temperature	$T_P = E_p/k = 1.417 \times 10^{32}$ K

References

[1] I. Newton, Philosophiae Naturalis Principia Mathematica, 1687.

[2] U. Le Verrier, Lettre de M. Le Verrier à M. Faye sur la théorie de Mercure et sur le mouvement du périhélie de cette planète, Comptes rendus hebdomadaires des séances de l'Académie des sciences (Paris) 49 (1859) 379.

[3] A. Einstein, Zur elektrodynamik bewegter körper, Annalen der Physik 17 (1905) 891.

[4] H. Minkowski, Raum und zeit, Physikalische Zeitschrift 10 (1908) 75.

[5] H.A. Lorentz, Electromagnetic phenomena in a system moving with any velocity smaller than that of light, in: KNAW Proceedings, 1903, p. 809.

[6] J.L. Lagrange, Mécanique Analytique, Courcier, 1811.

[7] E.B. Christoffel, Ueber die transformation der homogenen differentialausdrücke zweiten grades, Journal für die Reine und Angewandte Mathematik 70 (1869) 46.

[8] C.W. Misner, K.S. Thorne, J.A. Wheeler, Gravitation, Princeton University Press, 1973.

[9] A. Einstein, Die feldgleichungen der gravitation, Sitzungsberichte der Preussischen Akademie der Wissenschaften zu Berlin (1915) 844.

[10] A. Einstein, Die grundlage der allgemeinen relativitätstheorie, Annalen der Physik 354 (1916) 769.

[11] D. Hilbert, Die grundlagen der physik, Nachrichten Von der Gesellschaft der Wissenschaften Zu Göttingen, Mathematisch-Physikalische Klasse 3 (1915) 395.

[12] K. Schwarzschild, Uber das gravitationsfeld eines massenpunktes nach der einsteinschen theorie, Sitzungsberichte der Königlich Preussischen Akademie der Wissenschaften 7 (1916) 189.

[13] R.P. Kerr, Physical Review Letters 11 (1963) 237.

[14] G.D. Birkhoff, Relativity and Modern Physics, Harvard University Press, 1923.

[15] V.M. Slipher, Radial velocity observations of spiral nebulae, The Observatory 40 (1917) 304.

[16] A. Friedmann, Über die krümmung des raumes, Zeitschrift für Physik 10 (1922) 377.

[17] G. Lemaître, Un univers homogène de masse constante et de rayon croissant rendant compte de la vitesse radiale des nébuleuses extra-galactiques, Annales de la Société Scientifique de Bruxelles A 47 (1927) 49.

[18] E. Hubble, A relation between distance and radial velocity among extra-galactic nebulae, Proceedings of the National Academy of Sciences 15 (1929) 168.

[19] Planck Collaboration, Planck 2018 results – VI. Cosmological parameters, Astronomy & Astrophysics 641 (2020) A6.

[20] WMAP Collaboration, Wilkinson Microwave Anisotropy Probe: Overview, 2009.

[21] P.J. Steinhardt, N. Turok, A cyclic model of the universe, Science 296 (2002) 1436.

[22] R. Penrose, Before the big bang: an outrageous new perspective and its implications for particle physics, in: Proceedings of EPAC, Edinburgh, Scotland, 2006, p. 2759.

[23] M. Rauch, J. Miralda-Escudé, W.L.W. Sargent, T.A. Barlow, D.H. Weinberg, L. Hernquist, N. Katz, R. Cen, J.P. Ostriker, The opacity of the lyα forest and implications for ω_b and the ionizing background, The Astrophysical Journal 489 (1) (1997) 7.

[24] F. Zwicky, Die rotverschiebung von extragalaktischen nebeln, Helvetica Physica Acta 6 (1933) 110.

[25] H.W. Babcock, The rotation Andromeda nebula, Lick Observatory Bulletin 19 (1939) 41.

[26] V.C. Rubin, W.K. Ford Jr., Rotation of the Andromeda nebula from a spectroscopic survey of emission regions, The Astrophysical Journal 159 (1970) 379.

[27] M. Birkinshaw, The Sunyaev-Zel'dovich effect, Physics Reports 310 (1999) 97.

[28] D. Clowe, A. Gonzalez, M. Markevitch, Weak-lensing mass reconstruction of the interacting cluster 1e 0657-558: direct evidence for the existence of dark matter, The Astrophysical Journal 604 (2) (2004) 596.

[29] E. Salpeter, The luminosity function and stellar evolution, The Astrophysical Journal 121 (1955) 161.

[30] G.E. Miller, J.M. Scalo, The initial mass function and stellar birthrate in the solar neighborhood, The Astrophysical Journal. Supplement Series 41 (1979) 513.

[31] P. Kroupa, C. Weidner, J. Pflamm-Altenburg, I. Thies, J. Dabringhausen, M. Marks, T. Maschberger, The Stellar and Sub-Stellar Initial Mass Function of Simple and Composite Populations, Springer Netherlands, Dordrecht, 2013, pp. 115–242.
[32] S. Chandrasekhar, The highly collapsed configurations of a stellar mass (second paper), Monthly Notices of the Royal Astronomical Society 95 (1935) 207.
[33] B. Pacynski, Gravitational microlensing at large optical depth, The Astrophysical Journal 301 (1986) 503.
[34] P.A. Zyla, et al., Review of particle physics, Progress of Theoretical and Experimental Physics 2020 (8) (2020) 083C01.
[35] L. Calder, O. Lahav, Dark energy: back to Newton, Astronomy and Geophysics 48 (2008) 1.
[36] B. Moore, Evidence against dissipation-less dark matter from observations of galaxy haloes, Nature 370 (1994) 6491.
[37] M.L. Mateo, Dwarf galaxies of the local group, Annual Review of Astronomy and Astrophysics 36 (1998) 435.
[38] B. Moore, S. Ghigna, F. Governato, G. Lake, T. Quinn, J. Stadel, P. Tozzi, Dark matter substructure within galactic halos, The Astrophysical Journal 524 (1999) L19.
[39] A. Klypin, A.V. Kravtsov, O. Valenzuela, F. Prada, Where are the missing galactic satellites?, The Astrophysical Journal 522 (1) (1999) 82.
[40] S.Q. Hou, J.J. He, A. Parikh, D. Kahl, C.A. Bertulani, T. Kajino, G.J. Mathews, G. Zhao, Non-extensive statistics to the cosmological lithium problem, The Astrophysical Journal 834 (2) (2017) 165.
[41] L. Verde, T. Treu, A.G. Riess, Tensions between the early and late universe, Nature Astronomy 3 (2019) 891.
[42] A.G. Riess, W. Yuan, L.M. Macri, D. Scolnic, D. Brout, S. Casertano, D.O. Jones, Y. Murakami, G.S. Anand, L. Breuval, T.G. Brink, A.V. Filippenko, S. Hoffmann, S.W. Jha, W. D'arcy Kenworthy, J. Mackenty, B.E. Stahl, W. Zheng, A comprehensive measurement of the local value of the Hubble constant with 1 km s^{-1} Mpc^{-1} uncertainty from the Hubble Space Telescope and the SH0ES team, The Astrophysical Journal Letters 934 (1) (2022) L7.
[43] S. Torres-Flores, B. Epinat, P. Amram, H. Plana, C. Mendes de Oliveira, GHASP: an HI$\pm$ kinematic survey of spiral and irregular galaxies – IX. The near-infrared, stellar and baryonic Tully-Fisher relations, Monthly Notices of the Royal Astronomical Society 416 (3) (2011) 1936–1948.
[44] S.S. McGaugh, F. Lelli, J.M. Schombert, Radial acceleration relation in rotationally supported galaxies, Physical Review Letters 117 (20) (2016).
[45] S. Perlmutter, et al., Measurement of ω and λ from 42 high-redshift supernovae, The Astrophysical Journal 517 (1999) 565.
[46] A.G. Riess, et al., Observational evidence from supernovae for an accelerating universe and a cosmological constant, The Astronomical Journal 116 (1998) 1009.
[47] The Royal Swedish Academy of Sciences. Scientific background on the Nobel prize in physics 2011 the accelerating universe, https://www.nobelprize.org/uploads/2018/06/advanced-physicsprize2011.pdf, 2011.
[48] A.G. Riess, et al., New Hubble space telescope discoveries of Type Ia supernovae at $z \geq 1$: narrowing constraints on the early behavior of dark energy, The Astrophysical Journal 659 (1) (2007) 98.
[49] M. Kowalski, et al., Improved cosmological constraints from new, old, and combined supernova data sets, The Astrophysical Journal 686 (2) (2008) 749.
[50] A. Conley, et al., Supernova constraints and systematic uncertainties from the first three years of the supernova legacy survey, The Astrophysical Journal. Supplement Series 192 (1) (2010) 1.
[51] N. Suzuki, et al., The Hubble space telescope cluster supernova survey. V. Improving the dark-energy constraints above $z > 1$ and building an early-type-hosted supernova sample, The Astrophysical Journal 746 (1) (2012) 85.
[52] Planck Collaboration, Planck 2015 results – I. Overview of products and scientific results, Astronomy & Astrophysics 594 (2016) A1.
[53] W. Hu, M. White, Cosmic symphony, Scientific American 290N2 (2004) 44.
[54] D. McMahon, Relativity Demystified, McGraw Hill, 2005.
[55] B. Schutz, A First Course in General Relativity, Cambridge University Press, 2009.

[56] M. Dalarsson, N. Dalarsson, Tensors, Relativity and Cosmology, Academic Press, 2005.
[57] B. Ryden, Introduction to Cosmology, Cambridge University Press, 2002.
[58] D. Huterer, A Course in Cosmology: From Theory to Practice, Cambridge University Press, 2023.
[59] S. Weinberg, Cosmology, Oxford University Press, 2008.
[60] D. Hooper, Dark cosmos, in: Search of Universe's Missing Mass and Energy, Smithsonian, 2007.
[61] G. Bertone, D. Hooper, History of dark matter, Reviews of Modern Physics 90 (2018) 45002.
[62] L. Strigari, Galactic searches for dark matter, Physics Reports 531 (1) (2013).
[63] D. Merritt, A Philosophical Approach to MOND: Assessing the Milgromian Research Program in Cosmology, Cambridge University Press, 2020.
[64] S. Weinberg, The cosmological constant problem, Reviews of Modern Physics 61 (1) (1989).
[65] W. Hu, CMB tutorials, http://background.uchicago.edu, 1996.
[66] S. Dodelson, Modern Cosmology, Academic Press, 2003.

Index